MAGNETIC RECORDING

Hatschek

Technical Reviewers

Dr. Gordon Hughes, *Center for Magnetic Recording Research*
Dr. Richard Fayling, ret., *3M Corporation*
Dr. David L. Morton, Jr., *Rutgers, the State University of New Jersey*
Mr. J. G. McKnight, *Magnetic Reference Lab*
Dr. Roger Wood, *IBM Corporation*

Books of Related Interest from IEEE Press ...

THE STORY OF ELECTRICAL AND MAGNETIC MEASUREMENTS: From 500 B.C. to the 1920s
Joseph F. Keithley

1999	Hardcover	416 pp	IEEE Order No. PC5664	ISBN 0-7803-1193-0

MAGNETIC RECORDING TECHNOLOGY: Second Edition
C. Denis Mee and Eric D. Daniel

1996	Hardcover	750 pp	IEEE Order No. PC5659	ISBN 0-07-041276-6

MAGNETIC STORAGE HANDBOOK
C. Denis Mee and Eric D. Daniel

1996	Hardcover	752 pp	IEEE Order No. PC5688	ISBN 0-07-041275-8

MAGNETIC RECORDING
The First 100 Years

Edited by:

Eric D. Daniel
Formerly with Memorex Corporation
Santa Clara, California

C. Denis Mee
Formerly with IBM Corporation
San Jose, California

Mark H. Clark
Oregon Institute of Technology
Klamath Falls, Oregon

**IEEE
PRESS**

IEEE Magnetics Society, *Sponsor*

The Institute of Electrical and Electronics Engineers, Inc., New York

This book and other books may be purchased at a discount
from the publisher when ordered in bulk quantities. Contact:

IEEE Press Marketing
Attn: Special Sales
Piscataway, NJ 08855-1331
Fax: (732) 981-9334

For more information about IEEE Press products,
visit the IEEE Home Page: http://www.ieee.org

Printed in the United States of America

10 9 8 7 6 5 4 3 2 1

ISBN 0-7803-4709-9
IEEE order Number: PP5396

Library of Congress Cataloging-in-Publication Data

Magnetic Recording: The First 100 Years / edited by Eric D. Daniel, C. Denis
   Mee, Mark H. Clark.
        p.   cm.
     Includes bibliographical references and index.
     ISBN 0-7803-4709-9
     1. Magnetic recorders and recording--History.   I. Daniel, Eric D.
II. Mee, C. Denis.   III. Clark, Mark H., 1961–   .
TK7881.6.A13 1998
621.382'34—dc21   98-8207
                                                                        CIP

This book is dedicated to the memory of Dr. Hiroshi Sugaya,
the author of two chapters in this book and of
Helical-Scan Recorders for Broadcasting, and *Consumer Video Recorders.*

Hiroshi, a renowned authority on video recording, died in October 1997.
He will be missed by all his many colleagues and friends around the world.

Eric D. Daniel
C. Denis Mee
Mark H. Clark
January 1998

Contents

Chapter 4   Steel Tape and Wire Recorders   30
Mark H. Clark

Chapter 5   The Introduction of the Magnetophon   47
Friedrich K. Engel

VIDEO RECORDING

Chapter 9   The Challenge of Recording Video   124
Frederick M. Remley

Chapter 10   Early Fixed-Head Video Recorders   137
Finn Jorgensen

Chapter 11   The Ampex Quadruplex Recorders   153
John C. Mallinson

DATA RECORDING

Chapter 15   Capturing Data Magnetically   221

James E. Monson

Chapter 16   Data Storage on Drums   237

Sidney M. Rubens

Chapter 17   Data Storage on Tape   252

William B. Phillips

Chapter 18   Data Storage on Hard Magnetic Disks   270

Louis D. Stevens

Acknowledgments

In putting together a book of this kind, the editors have had the cooperation and help of many people. In the first place, we thank and applaud the 16 authors. Collectively, they represent a pool of expertise that covers the spectrum of magnetic recording. Many of them were there when history was made; indeed, some were the pioneers and inventors who made history in their day. We are grateful that they took time away from busy schedules, or interrupted their retirement, to research and write their contributions.

We also wish to acknowledge the assistance given by many others in encouraging this book to be written, in making suggestions about its scope, in reviewing and suggesting improvements in the various chapters, and in tracking down illustrations. In this regard, we owe thanks to James Gandy, Beverley Gooch, Pete Hammar, Roger Hoyt, Gordon Hughes, Steven Luitjens, Jay McKnight, Jack Mullin, Masaaki Notani, Lou Ottens, Fred Remley, and Roger Wood. Special thanks are owed to Jim Monson, who took part in early planning and helped out in some of the more difficult tasks. We also thank various persons and organizations for providing illustrations from their archives, including AEG, Ampex, BASF, BBC, Bing Crosby Enterprises, the Danish Museum of Science and Technology, Datatape, ERA, IBM, Matsushita, Jack Mullin, NHK, the Oberlin Smith Society, Odetics, the Peder O. Pedersen Archive, Philips, Sony, and Univac.

Eric D. Daniel
C. Denis Mee
Mark H. Clark

Contributors

Mark H. Clark *Oregon Institute of Technology,* Klamath Falls, Oregon, U.S.A.

Friedrich K. Engel Weinheim-Hohensachsen, Germany; formerly with *BASF,* Germany.

Beverley R. Gooch *Ampex Corporation,* Redwood City, California, USA.

Finn Jorgensen *Danvik,* Santa Barbara, California, USA.

John C. Mallinson *Mallinson Magnetics,* Belmont, California, USA.

James E. Monson *Harvey Mudd College,* Claremont, California, USA.

Henry Nielsen *Institute for History of Science, University of Aarhus,* Denmark.

David L. Noble Monte Sereno, California, USA; formerly with *IBM.*

William B. Phillips Tucson, Arizona, USA; formerly with *Storage Tek.*

Frederick M. Remley Ann Arbor, Michigan, USA; formerly with the *University of Michigan.*

Sidney M. Rubens Saint Paul, Minnesota, USA; formerly with *Univac.*

Koichi Sadashige Voorhees, New Jersey, USA; formerly with *RCA.*

Louis D. Stevens Aptos, California, USA; formerly with *IBM.*

Hiroshi Sugaya Suita, Osaka, Japan; formerly with *Matsushita.*

John R. Watkinson *Watkinson International Communications,* Reading, Berkshire, England

Figure 1 Oberlin Smith, 1840–1926, was the first to conceive and describe the principles of magnetic recording. (Photograph courtesy of the Oberlin Smith Society.)

Figure 2 Valdemar Poulsen, 1869–1942, was the first to demonstrate a working model of a magnetic recorder, and obtain a patent on his invention. (Photograph courtesy of Peder O. Pedersen Archive.)

1 Introduction

C. Denis Mee and Eric D. Daniel

The principles of magnetic recording were described admirably by Oberlin Smith (shown in Figure 1, left) in 1878. But it was not until 1898 that a device using these principles was demonstrated and patented by the Danish inventor Valdemar Poulsen (Figure 2, left). The publication of this book in 1998 is intended both to honor Smith and to celebrate the centenary of Poulsen's invention.

THE GROWTH OF MAGNETIC RECORDING

After a slow start—it was 30 years after Poulsen's invention before modest commercial success was achieved—magnetic recording usage expanded rapidly, and today it enters into almost every facet of our working and leisure activities. Magnetic tape recorders remain the principal means of recording music in professional studios, the home, and the automobile. Broadcasters use magnetic recording to provide our television shows; we use it to record video programs, play movies, and take video camera shots of our children. Digital still cameras using magnetic recording are increasingly popular. Magnetic recording is an essential ingredient in mainframe computers, as well as in personal computers used at the office and at home. The magnetic stripes on credit cards and cinematographic film are other examples of the ubiquitous roles of magnetic recording in our lives.

Most people are aware that these technical marvels exist. Strangely, many do not fully realize that magnetic recording lies at the root of them all. The words "magnetic recording" still tend to conjure up the image of an open-reel sound recorder of the Nixon era, particularly to members of the older generation. The words are seldom linked straight off with hard disk storage, camcorders, or one of the many other miniaturized devices that now abound, in which the recording medium is hidden from view.

Worldwide, the magnetic recording industry now has annual revenues of over $100 billion—not all that far behind the semiconductor business. Magnetic recording employs over 500,000 people distributed throughout the industrialized world. Most developed countries have a significant level of activity, and manufacturing facilities exist in many less-developed countries.

If Valdemar Poulsen or Oberlin Smith were alive today, they would be amazed—perhaps more than a little frightened—to see how their pioneer work has been transformed into this varied and ubiquitous technology. Our object in putting this book together is to describe how this transition took place and, for the first time, to do so in a comprehensive way that treats all the main application areas: audio (sound) recording, video (television) recording, and data (computer and instrumentation) recording.

THE FOUR ENDURING PRODUCT FORMATS

A word about the organization of the book. After making outlines of several approaches, we decided that the best way to maintain continuity would be to treat the main application areas separately and, within each area, to deal with the development of significant new products, or technologies, in approximate chronological order. Accordingly, the first chapters are devoted to audio recording, since this was all there was from 1898 until about 1945. The middle chapters cover video recording, which had its start in the early 1950s. The last chapters deal with data recording, which also got going in a significant way in the early 1950s.

Over the years of its existence, each area expanded in its applications, and product designs have proliferated. Some design formats were destined to be short-lived; others lasted for many years before being abandoned or replaced. Examples of extinct formats are those based on media in the form of wire, drums, and various kinds of strips, loops, sheets, and cards.

Only four generic design formats have shown long-term endurance. Equally remarkable is that these formats were established by four pioneer product developments made 30 to 60 years ago: one in audio, one in video, and two in data recording. The four developments were:

- *The Magnetophon audio recorder.* This pioneer development by I.G. Farben and BASF in Germany, in the 1930s and 1940s, introduced recorders using magnetically coated plastic tape and ring heads. Once the development became widely known, it rapidly replaced its crude predecessors based on steel wire and tape. The Magnetophon established the generic format for all subsequent magnetic tape recorders using stationary heads.

- *The quadruplex video recorder.* This development by Ampex, in 1956, introduced a tape recorder using four heads mounted on a rotating wheel. The recorder met the needs of a rapidly expanding television industry for a convenient replacement of photographic recording of television programs. The quadruplex established the generic format for all video recorders, in which heads are scanned at very high speed over a relatively slowly moving tape.

- *The RAMAC* disk file.* This development by IBM, also in 1956, introduced fast-access data recorders using coated rigid disks, and heads having random access to any recorded track. The development followed the introduction of electronic computers, and it became urgent to provide data storage devices having much faster access than tape. The RAMAC established the generic format for all "hard" disk drives, in which heads fly close to the surface of a disk rotating at high speed and can be moved randomly to any radial position.

- *The diskette.* This development by IBM, in the late 1960s, introduced data recorders using coated flexible disks, and heads having random access to any recorded track. Originally designed to fill a need for a means of loading microcode into a computer, the diskette soon became widely used as an inexpensive, removable storage device. The diskette established the generic format for all subsequent "floppy" disk drives, in which heads are in contact with the surface of a disk and can be moved randomly to any radial position. The disks are contained in a protective jacket to allow safe removal.

The four formats—two tape-based, two disk-based—set the stage for all the successful, high-technology, magnetic recording products that exist today.

After its introduction, a preferred format flourished within its application area, nourished by a continual flow of innovations in the design of recording components such as media and heads, and by the introduction of new technologies, such as the transistor and large-scale integration. For example, one can trace the development of audio tape recorders from the Magnetophon all the way to the small, convenient cassette recorders that exist today. One can follow the steps that transformed the bulky Quadruplex recorder, with its 2-inch-wide tape, into today's 8 mm video cassette recorders. Similarly, the huge RAMAC led eventually to the hard-disk drives, with heads flying at one-millionth of an inch over a 3.5-inch (or smaller) sputtered metal disk, which are at the heart of today's personal computers. Finally, the 8-inch diskette evolved into the now universally used 3.5-inch floppy disk in its hard jacket, and several more expensive, high-capacity disks are available.

---

*Random access method of accounting and control.

Another trend gave life to the preferred formats as well: they soon spawned products outside their original application area. Thus, the audio tape recorder format moved outside the audio realm to form the basis for the design of computer and instrumentation tape drives. Later, the video tape recorder format was also adapted for data recording purposes, and for digital audio applications. The hard disk format took a little longer to be put to use outside the computer area, but eventually found a place in audio and video editing, and as a "jukebox" for commercials. The floppy disk format has found a place in still photography.

A couple of examples may help to illustrate the progress made over the years. We begin with an audiovideo application—the first concert recorded on magnetic tape, which was a program of Mozart conducted by Sir Thomas Beecham in Berlin in 1936. A reasonable guess suggests that one hour's recording of this concert with the equipment then available took about 5000 meters of 6.3 mm tape, or about 25 large reels. The quality was very poor and, of course, the recording itself was monaural. Today, the same concert could be recorded in high-quality stereo, together with a color video recording of the orchestra, on a single 8 mm video cassette.

Another example is taken from the computer field. RAMAC, the first hard disk storage device, occupied a space of about 1 million cubic centimeters (1 cubic meter), and it achieved a storage density of about 3 bits per square millimeter. A modern disk drive occupies less than 100 cm^3 and stores data at a density of more than 4 million bits/mm^2. On a bits-per-unit-volume basis, the improvement is quite spectacular—a factor of over 10 million. Today, the hard disk drive has the highest areal density of any magnetic recording device.

A LOOK AT THE FUTURE

Forecasting the future is always risky. We can, however, make some predictions without too much danger of looking foolish down the road. First it is evident that signals will be increasingly recorded in digital form. The technology for doing this has been developed to a substantial degree, and conversion to digital has already occurred in many professional areas. The acceptance of digital magnetic recording by consumers has been slow, but acceptance may widen when prices are reduced and certain copyright problems are resolved.

The second trend that can be predicted is that the density of data storage on magnetic media of all types will continue to increase. In particular, the increase in the areal density of hard disks over the last 10 years has been truly phenomenal, amounting to over 50% annually. This, in conjunction with smaller, lower cost drives, has driven the cost per megabyte down to 5 cents compared with $10 a decade ago. The decline cannot continue indefinitely, but there is no sign of any immediate slowdown. The increase in the storage capacity of floppy disks has also been remarkable, and further enhancements can be expected.

Nevertheless, some fundamental limitations to the indefinite increase of magnetic recording density are looming. Before long, the size of a recorded bit will reach a point at which, theoretically, it is no longer magnetically stable. Also, decreasing the bit size implies an increase in the data rate to a point that causes the heads to run into frequency problems. Judging from past history, however, ways will be found to avoid or overcome these barriers. For example, some combination of magnetic recording (high bit density) and optical recording (high track density) is a possibility.

The third prediction is that competition to magnetic recording from optical disk recording will continue to occur in certain areas. So far optical recording has been successfully introduced in three read-only formats which, in the order of their appearance, are the audio *compact disc* or *CD*, the *CD-ROM* , and the *digital versatile disk* or *DVD*. Owing to its high sound quality and convenience, the CD now dominates the prerecorded music business, and the *CD-ROM* is a ubiquitous element of personal computers. The *DVD* is designed to be used both in computers and in the prerecorded video business, where it could have a large impact once software has become available and the price of equipment has dropped. Upon production of commercially viable versions of the optical disk recorders that can write and erase, as well as read, these devices may replace magnetic recording for consumer use. In computer applications, the optical disk has offered no real threat to the hard disk, and a limited direct threat to the floppy. Overall, however, the prime attributes of magnetic disk storage, and their continuing improvement, will provide a difficult target for optical storage to challenge across the board.

2 The Magnetic Recording of Sound

Mark H. Clark

Within the brief span of two years, three inventions central to the history of recorded sound were born. At the great Centennial Exposition of 1876 in Philadelphia, Alexander Graham Bell unveiled the telephone, which translated the vibrations of sound into electrical current sent along a wire, then translated the current back into sound again. The first device of its kind, the telephone captured the headlines of newspapers around the world, and made its inventor famous. The ancestor of all microphones and loudspeakers, Bell's telephone and the industry it spawned have played a crucial role in shaping the way we listen to the world around us.

In the following year, 1877, the idea of recording and reproducing sound mechanically was proposed by Charles Cros in France. Later that year, Thomas Edison in the United States reduced the idea to practice with his *phonograph*. Edison's recording medium consisted of a grooved metal cylinder over which was stretched a sheet of tinfoil. Sound impinging against a diaphragm caused a stylus to press into the surface of the tinfoil, embossing it, while the cylinder was cranked by hand. On playback, the tinfoil caused the stylus to vibrate the diaphragm to reproduce the sound. Edison was already known as an innovator, but the phonograph increased his fame, as the public wondered at this new marvel. The first machine that could record sound and reproduce it later, the phonograph forever changed the relationship between entertainers and their public. No longer did a singer and an audience have to meet in the same place and time—a voice could be multiplied and heard anywhere there was a phonograph. Artists could—and did—make money without leaving home.

The third invention was that of magnetic recording in 1878. The inventor, Oberlin Smith, did not obtain a patent, nor did he become famous. As far as we know, he failed to make a working model, and therefore his invention was never demonstrated to the public or heralded in the press. However, Smith's ideas were sound, his description of them is delightfully lucid, and, on occasion, his editorial comments are amusingly prophetic. There is no better way of introducing the reader to the basic principles of magnetic recording than to quote from his writings.

OBERLIN SMITH'S IDEA

The conception of magnetic recording came hard on the heels of the invention of the phonograph. In early 1878, Oberlin Smith, mechanical engineer and founder of the Ferracute Machine Company, visited Thomas Edison's laboratory in Menlo Park, New Jersey. Edison had just patented his cylinder phonograph, and Smith found this new invention fascinating. Upon returning to his home in Bridgeton, New Jersey, he began to experiment with ways to improve Edison's invention.

Smith was inspired by what he thought was a significant flaw in the phonograph. Mechanical recording inevitably introduced noise into the reproduction process, through the friction between the playback needle and the recorded groove. Since mechanical contact was essential to produce the sound, the noise could not be eliminated, only reduced. Smith thought of magnetic recording as a means of avoiding this problem, since no actual physical contact would be necessary.

On September 23, 1878, Smith recorded his ideas in a memorandum, reproduced below, in which he described for the first time the concept of magnetic recording. A version of Smith's schematic is given in Figure 2–1.

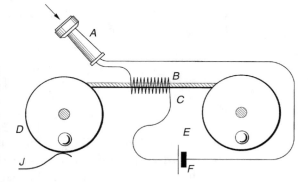

Figure 2–1 Drawing from Smith's 1888 article in *Electrical World*. The machine is shown ready to record; *A* is the microphone, *B* is the recording coil, *C* is the wire or string that serves as the recording medium, *D,E* are the transport spools, *F* is the battery, and *J* is a tension brake to keep *C* taut. [*Courtesy of Friedrich Karl Engel.*]

Bridgeton, N. Jersey, Sept. 23, 1878.
MEMORANDUM:-

I have invented another improvement in talking phonographs, and described it to Mr. Fred F. Smith. (That's so, Fred F. Smith). It is, however, subject to experimental investigation of the capability of a small
wire (probably tempered steel) to receive magnetism (by induction
from an electrical current in a short surrounding helix) in spots or
zones of varying intensity at different portions of its length.

Assuming this capability, my "receiver", or "listener", is as follows: - A mouthpiece, diaphragm, carbon button &c. of an Edison telephone. Reels for moving a wire through a helix and a battery.

While talking into the mouthpiece, the varying intensity of current (caused by the varying condensation of the carbon) produces
zones, or spots, of magnetism in the wire which vary in length and
strength in accordance with the length and amplitude of the sound vibrations. The wire becomes the record of the voice, instead of tinfoil.
The "talker" consists of the same reels; the same or another helix, battery &c.; a Hughes microphone and a Bell telephone. The magnetic
wire, being passed through the helix, induces a delicate series of currents of magnetoelectricity which pass through the microphone and
are given out as sound vibrations by the telephone: or *otherwise*.

The advantages in cheapness, simplicity and delicacy, are mani-
fest——also the facility with which the record may be kept wound on
cheap spools, like sewing cotton.

[signed] Oberlin Smith.

If wire E will not magnetize in "spots", it could be a *chain* of alternate
links of steels & non-magnetizable material.

Sep. 24,
F.F. Smith & I have just jointly suggested a wire, made of brass, lead
or other metal, impregnated mechanically with steel dust——probably
hardened in the wire.

0. Smith.

This was described to me Sept. 24th, 1878.

P. K. Reeves.

The memorandum is striking in that it describes all the elements of a modern
magnetic recorder, including a recording medium, an elementary record/repro-
duce transducer, and a transport mechanism. Smith even notes the possibility of
using a nonuniform recording medium, such as a brass wire or cotton string

impregnated with iron particles. The only really weak element was his use of a simple air-cored helix as a record/reproduce transducer. There is, however evidence that he considered a more practical type of "head" using magnetic pole pieces (Fig. 2–2).

Unfortunately Smith's laboratory notes no longer exist (two separate fires destroyed most of his personal papers), so we do not know the exact nature of the inventor's development work. However, based on his published writings and a few surviving drawings, we know that he did build several pieces of experimental apparatus and that he carried out a series of trials. We know that he built at least one and possibly more reel-to-reel mechanisms to test his ideas. He obtained a "button" of mercury-impregnated carbon from Edison late in December 1878, and apparently used it as a microphone in his recording trials. Finally, he built a machine that spun magnetic powder onto cotton thread, which he planned to use as a recording medium. According to Smith, he was never able to obtain "acoustic" results, but the historian Friedrich Karl Engel, who has made a careful study of the relevant documents, believes that Smith did in fact have at least limited experimental success.

Smith abandoned his experiments sometime between December 1878 and September 1888, probably in early 1879. The growing demand for the stamping presses his company Ferracute produced left him no choice but to focus on more practical matters.

In September 1888, in response to the "great activity of thought regarding listening and talking machines," stemming from the renewed interest in the phonograph and gramophone of Edison and others, Smith published an account of his work in the journal *Electrical World*. The article was also translated and published in the French publication *La Lumière Electrique* later in 1888. The article, which describes in greater detail the principles set forth in his memorandum of 1878, contains many gems of insight and prediction. For example, he thought that demagnetization (for which he used the wonderful term "magnetic depravity") might be a problem, particularly in a steel wire medium. He was also aware that the reproduced currents might not be sufficient, but might need an

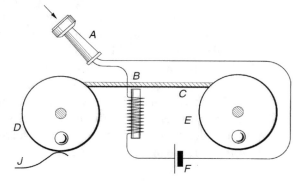

Figure 2–2 Drawing reconstructing alternate recording method suggested by Smith in letter to *Electrical World;* components are labeled as in Figure 2–1: *B* is an electromagnet pressed against the recording wire or string and perpendicular to it. [*Courtesy of Friedrich Karl Engel.*]

"intensifying apparatus," something that was not to be "thought out" until Lee de Forest produced his Audion amplifier some 20 years later.

It is worthwhile quoting the last paragraph of Smith's article in full:

> [The writer] has not the time, to say nothing of a properly equipped laboratory, to carry out the ideas suggested to their logical conclusion of success or failure, and, therefore, makes them public, hoping that some of the numerous experimenters now working in this field may find in them a germ of good from which something useful may grow. Should this be the case, he will doubtless get due credit for his share in the matter; but if, on the other hand, these suggestions prove worthless, they will still have served a purpose, on the principle that a demonstration of what *can't* be done is often a pertinent hint as to what *can* be.

Oberlin Smith certainly should get "due credit for his share in the matter." No historical review of magnetic recording would be complete without praising his contribution.

MAKING MAGNETIC RECORDING WORK

We have used Oberlin Smith to introduce the basic ideas behind magnetic recording. In the next chapter we go to a more practical level, describing the work of Valdemar Poulsen, who is generally regarded as the true inventor of magnetic recording in that, in 1898, he obtained the first patent and produced the first working machine, the Telegraphone. We will find, however, that numerous attempts to launch the Telegraphone commercially failed, and the effort was abandoned in 1918.

When the principles were well understood, and a working model had been demonstrated, why did the magnetic recording of sound have such a disappointing first 20 years? To answer this question, we need to appreciate that magnetic recording is only a tool for storing information—nothing more. To make that information useful and available, additional devices are required to link the recorder with the outside world.

In the first place, devices are needed to translate the vibrations that constitute sound into electrical signals, and vice versa. Very early magnetic recorders borrowed directly from the telephone industry, using the transmitters and receivers pioneered by Bell and Edison. It is no accident that Valdemar Poulsen was a telephone technician, nor that telephone companies would play a major role in the early years of magnetic recording's development. Many years passed before more sophisticated microphones and loudspeakers were developed.

The second major enabling technology was electronic amplification. The first magnetic recorders were developed and built before any practical method existed for amplifying electrical signals. The voltages induced in the playback heads were weak, and the sounds they produced in the earphones were barely audible. The coming of the vacuum tube amplifier in the 1910s and 1920s was a great boon for magnetic recording. With electronic amplification, the sounds that had been so faint could be made distinct, and the magnetic recorder could at last compete with the phonograph in terms of volume of sound.

BIASING TECHNIQUES

The human ear is exquisitely sensitive to two defects that are inevitably present to some extent in any reproduced sound: noise—the presence of hiss or crackles that were not in the original sound; and distortion—the introduction of jarring, unwanted sounds and a lack of clarity at the higher levels. Early magnetic recorders had a lot of both. It was soon found that adding a constant current to the signal current in the recording head resulted in much lower distortion. This technique, known as direct current bias (dc bias), became standard on all magnetic recorders produced prior to the early 1940s, and on some after that. The use of dc bias did nothing to reduce the noise; in fact, it made it worse.

Even with dc bias, the clarity of the recorded signal on playback was not outstanding, and was inferior to that of the best phonographs of the 1920s and 1930s, particularly at high frequencies. In the late 1930s, however, researchers in different locations rediscovered a technique that had been proposed in 1920. They found that replacing the direct current bias with a comparable level of alternating current, having a frequency well above the highest frequency of the sound signal, had a dramatic effect on the quality of magnetic recordings. Both distortion and noise were greatly reduced, leading to reproduced sounds having a fidelity equal to or better than that offered by the competition. Ac bias, as the technique came to be known, paved the way for much wider applications of audio magnetic recording by making it possible to build machines of higher fidelity for the same cost, or equal fidelity at much lower cost.

MEDIA AND HEAD DESIGN

Although the recording process seemed straightforward, optimal designs for the critical components of a recorder proved to be difficult to produce. It took a long time to find inexpensive and reliable recording media, and the design of the recording and reproducing heads went through many changes before settling on an optimal basic configuration.

In the early years of magnetic recording, the standard recording medium was steel in the form of wire (piano wire) or thin ribbon. Although it was possible to make fairly good quality recordings in this way, steel media suffered from some major disadvantages. An inherent disadvantage of wire is that it can twist between record and playback, producing severe fluctuations in performance at high frequencies. A steel ribbon does not have this problem, but it is bulky, and it proved to be expensive to ensure that this medium would have sufficiently uniform magnetic properties. Both wire and ribbon steel media shared a more fundamental disadvantage: the basic incompatibility between a steel having sufficient hardness and rigidity to impart good magnetic properties and the need for a medium flexible enough to be easily wound on a reel for storage.

It is for these reasons that the invention of coated media for magnetic recording was such a major advance. Although the idea of a coated medium is as old as the invention of magnetic recording itself, the first commercial coated-media recorder was not developed until the early 1930s in Germany. The widespread use of coated media was delayed by the second world war, but after 1945 this new recording method rapidly replaced solid steel media for almost all applications.

By separating the recording media into two parts—a magnetic material and a carrier that supported it—it became possible to optimize each part for its function. The magnetic properties of the recording part of the system were not limited by the need for mechanical strength, and the support could be made without concern for its magnetic properties. That the resulting product could be produced more cheaply than solid steel media was a nice bonus.

From the time of the development of ac bias and coated media in the late 1930s until recently, the technological path of magnetic recording was one of evolution rather than revolution. As described in Chapters 3 to 7, improvements to recording tape, recording heads, transport mechanisms, and supporting electronics were simply that: improvements. A great deal of development occurred but, at heart, the machines were identical in principle.

Beginning in the 1970s, however, a new technology revolutionized audio magnetic recording. That new technology was digital recording, which operates on a different principle. Rather than record sound as an analog of the waveform involved, digital recording first describes that waveform numerically, and then records the resulting numbers in digital form. As outlined in Chapter 8, this new recording method takes advantage of recent developments in microprocessor technology to perform the large volume of calculations necessary to accurately describe the waveform. The two most significant advantages of digital over analog recording are that, in principle, noise and other defects present in analog disappear, and that copies are exactly identical to originals—duplication and editing can be carried out without loss of sound quality.

APPLICATIONS

All this technology would be of merely academic interest if it had never been used. Who, then, are the users of audio magnetic recorders? In general, audio magnetic recorders have had three major applications. These can be loosely described as professional, business, and consumer.

In professional applications, audio magnetic recorders have found their primary application in the entertainment industry. This includes broadcasting, motion picture sound, and the creation of studio recordings for singers and instrumentalists. The machines for these applications have tended to be expensive, state-of-the-art devices that need skilled operators to extract optimum performance. Such operators tend to be more tolerant of complexity, particularly in routine use. In recent years digital machines have replaced analog ones in 50 to 70% of professional applications.

In business applications, audio magnetic recording has tended to replace existing technology and, in many cases, people. Answering machines replaced telephone receptionists, and tape recorders replaced stenographers for recording meetings or the conversations captured by police wiretaps. In the public service sector, magnetic recording is used to log such things as emergency calls. In general, these applications demand reliable, low-maintenance machines, although modest sound quality is acceptable.

Consumer audio magnetic recording differs primarily in that its focus has been almost exclusively on the reproduction of music. For the vast majority of consumers, ease of use has been an overriding consideration, although good sound quality is expected by the more sophisticated users. Cassette and cartridge recorders developed in the 1960s were largely responsible for the enormous growth in this market, since they made tape much easier to load and unload from machines.

The coming of digital recording has to some extent reduced the differences between these categories, although significant differences still exist between professional and consumer equipment. The quality of digital recording, especially in copying, makes it increasingly easy to achieve professional-style results with machines that are as cheap and easy to use as traditional consumer-level equipment.

IMPACT

What, then, has been the impact of audio magnetic recording on society? During early years of the industry the answer is simple—not much. Prior to the second world war, magnetic recording occupied a few niche markets, making the lives of a few radio personnel and military men a little easier. In the immediate postwar period, tape recording was rapidly accepted by the radio industry as a means of producing and rebroadcasting programs, including news; it also played an increasing role in sound for movies. More and more musicians and hi-fi fans used magnetic recorders, and businesses turned increasingly to tape recorders for dictation and telephone answering machines. Even so, the life of the average person was only indirectly changed by audio magnetic recording.

Beginning in the mid-1950s, however, two innovations fundamentally changed how people interact with music and recorded sound. The first change took place within the studio. Building on earlier work on stereo recorders, manufacturers began to turn out studio recorders that were capable of excellent sound quality, and they introduced multitrack studio recorders that allowed far greater manipulation of sound in the studio than previously. By combining multiple takes of a song to create a more perfect version, or by allowing a single musician to play all the instruments in a complex arrangement, multitrack recorders gave artists far greater control over the music they produced.

Was the music produced better than before? Perhaps. Perhaps not. What is certain is that the music was different, and the relationship between artist and public was fundamentally changed. No longer was the recording process an attempt to capture how an artist played live—increasingly, live performance became an attempt to duplicate what had been produced in the studio.

The second major change brought about by the magnetic recording industry was the development of easy-to-use portable systems. In the 1960s, the 8-track cartridge and the compact cassette provided consumers with the ability to take music on the road or to the beach. Two decades later, the Sony Walkman and its imitators made personal choice in musical programming even easier. Unlike the transistor radio, which tied the listener into a central broadcast and set playlist, the Walkman offered the freedom to record and play back only the songs one wanted to hear, when one wanted to hear them.

This, then, is the legacy of audio magnetic recording—music shaped in the studio by artists to their own exacting specifications, and sold to a public who can take that music anywhere, enjoy it in privacy, or share it with everyone nearby. Artistic expression, high quality, and personal choice—what more can one ask of any technology?

REFERENCES

Cox , Arthur J., and Thomas Malim, *Ferracute: The History of an American Enterprise,* Cowan Printing, Bridgeton, NJ, 1985.

Engel, Friedrich Karl, *Oberlin Smith and the Invention of Magnetic Sound Recording: An Appreciation on the 150th Anniversary of the Inventor's Birth,* Oberlin Smith Society, Bridgeton, NJ, 1990.

Smith, Oberlin, "Some Possible Forms of Phonograph," *The Electrical World,* 116, September 1888.

3 The Telegraphone

Mark H. Clark and Henry Nielsen

INVENTING THE TELEGRAPHONE

The first functioning magnetic recorder was built in 1898 by Valdemar Poulsen, a Danish engineer, 20 years after Oberlin Smith had come up with the idea. Poulsen was born in Copenhagen, the son of a lawyer who eventually became a judge on the Danish Supreme Court. Unable to pass the entrance examination at the Technical University of Denmark, the younger Poulsen was hired in 1893 as an assistant to the chief telephone engineer at the Copenhagen Telephone Company (KTAS) through his father's influence. In the late 1890s, Poulsen worked in the technical department of KTAS and also had regular contact with researchers at the Great Northern Telegraph Company. Poulsen worked as a troubleshooter and became familiar with the state of the art in both telegraph and telephone technology.

How did Poulsen grow interested in magnetic recording? As he explained later, he was frustrated by the inability of telephone users to leave a message when the party they called was not at home. Using telephone components, Poulsen experimented with the application of magnetism to the recording of telephone messages. In the summer of 1898, he discovered that he could record sound using a telephone microphone as an electrical current source which fed through an electromagnet. He was able to make short recordings by moving the electromagnet along a piece of piano wire while talking into the microphone. He could then play back the sound by connecting the electromagnet to a telephone earpiece and moving the electromagnet along the wire at the same speed and in the same direction used during the recording process (Fig. 3–1). Although crude, the device worked, and Poulsen realized that he had discovered a fundamentally new way of recording sound.

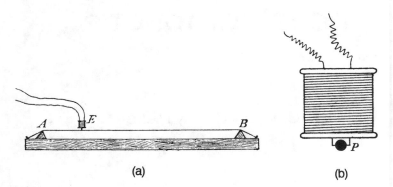

Figure 3–1   The apparatus Poulsen used for his first experiments. **(a)** A steel wire stretched between supports *A* and *B;* the electromagnet *E* was moved along the wire by hand to record or replay sound. **(b)** A close-up view of the electromagnet, oriented at right angles to Figure 3–1**a**: the solid dot *P* is a cross section of the steel wire, partially surrounded by the pole piece of the electromagnet. [*Courtesy of the Peder O. Pedersen Archive.*]

By the time of his first patent application on December 1, 1898, Poulsen had designed a more sophisticated machine that looked much like Edison's first phonograph. It recorded sound well enough to permit Poulsen to use it to convince Danish patent authorities and financial backers that he had a working device. He filed a patent application in Germany on December 9, 1898, eight days after the Danish application. In 1899 he left his job with the telephone company to devote full time to his recorder, which he had christened the "Telegraphone." Between March and May of 1899 Poulsen filed in 14 more European countries, and on June 7 he applied for an American patent. In those applications, Poulsen included drawings of his phonograph-like recorder (Fig. 3–2) and made broad claims related to recording sound magnetically. The text of the Danish application is typical:

> The object of the said invention is a Telegraphone, that is an electrical phonograph which will find its main applications in connection with the equipment employed in modern telephony. The Telegraphone can be used locally but is especially useful for distance-phonography, that is recording, storing and replaying speech as delivered from a standard telephone, as the Telegraphone in the absence of the called subscriber can substitute for the receiver telephone.

Poulsen was eventually able to obtain broad protection of the principle of magnetic recording in 38 nations. The subsequent commercial failure of Poulsen's invention was not due to inadequate patents; complete coverage in all likely markets was obtained, and the patents were never challenged in court.

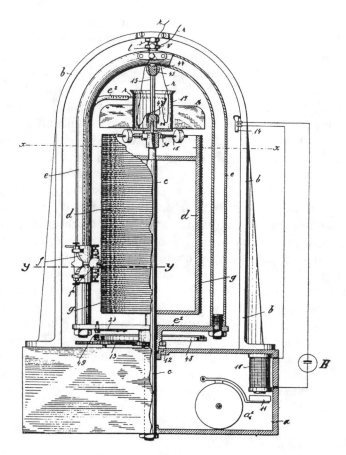

Figure 3–2　Drawing of the Telegraphone from the American patent application (U.S. Patent 661,619). The vertical cylinder was covered with a wire spiral, which acts as the recording medium. The recording head moved around the cylinder.

RESEARCH, DEVELOPMENT, AND THE GERMAN PARTNERSHIPS

Even before the initial Danish patent was granted, efforts were made to exploit the Telegraphone commercially. In the summer of 1899, Poulsen and a group of Danish investors established Aktieselskabet Telegrafonen Patent Poulsen, a Danish corporation. The company took over Poulsen's patent rights in exchange for shares in the firm, and Poulsen, having resigned from his job with the Copenhagen Telephone Company, became the new firm's director. By the end of 1899, the staff was expanded to include two engineers, P. O. Pedersen and E. S. Hagemann, the electrical technician J. P. Christensen, and the skilled machinist E. Lübcke. The latter two individuals were in charge of actually building experimental apparatus for the three engineers, while Hagemann was an assistant to Pedersen and Poulsen.

The board of Aktieselskabet Telegrafonen, headed by Lemvig Fog, soon found that commercializing the Telegraphone would require additional help. Although Poulsen thought he and his small technical staff could develop the device into a marketable product, the board, wanting a quick return on their investment, pushed for accelerated development. As a consequence of tariff laws that favored the import of finished products over the import of raw materials, the Danish electrotechnical industry was very small, and most electrical equipment was imported from Denmark's great southern neighbor, Germany. It is therefore not surprising that the board selected a German firm, Mix & Genest, as a partner.

A major manufacturer of telephone and telegraph equipment, Mix & Genest supplied telephone equipment to the Copenhagen Telephone Company. Thus Poulsen would have been familiar with the products. Moreover, Mix & Genest had its own research laboratory. The German firm seemed to be an ideal partner for further development work on the Telegraphone.

An informal agreement was reached in the winter of 1899, and as a result most of Aktieselskabet Telegrafonen's technical staff went to Berlin to work at the Mix & Genest research and development department. Their initial task was to design and build machines for exhibit at the Paris World Exhibition in June 1900, only a month after their arrival in Germany.

THE PARIS EXHIBITION

Displaying their invention at a world exhibition was a logical strategy for Poulsen and his collaborators. Ever since the Crystal Palace Exposition of 1851, world exhibitions had provided venues for the exposure of innovative new technologies to the widest possible audience. Poulsen was no doubt aware of the favorable publicity that could be generated. Active as he was in the telephone industry, he surely would have been familiar with Alexander Graham Bell's dramatic demonstration at the American Centennial Exposition in Philadelphia in 1876. More recently, Thomas Edison had shown off his electric light at the 1881 Paris Exhibition and his phonograph at the 1889 Paris celebration of the centennial of the French Revolution. George Westinghouse had demonstrated his new alternating current electrical power system to wide acclaim at the Columbian Exposition of 1893. No doubt Poulsen and his collaborators hoped the Telegraphone would achieve similar recognition.

In preparation for their debut in Paris, Poulsen, Pedersen, Christensen, and Lübcke stayed in Berlin until early June, working daily with Mix & Genest's engineers and technicians, most notably H. Zopke and Ernst Ruhmer. They worked to improve the existing cylinder Telegraphones and also initiated the development of several new designs in an attempt to improve sound clarity and volume, including reel-to-reel machines using steel tape or wire to replace the old cylinder model. Innovations included the use of different wire or tape dimensions, different recording head configurations, and variations in transport speed.

In the course of these experiments the team experienced at least one definite success, the discovery that a recording can be substantially improved by means of dc biasing. The discovery was made on May 14 or 15, 1900, during a series of systematic attempts to improve the quality of sound recorded on a new steel tape machine. In each run a brief message was recorded on the tape, then replayed, judged for quality, and finally erased by a strong magnet. This left the tape, or wire, in a state of maximum magnetization (i.e., "saturated") in one direction. What Poulsen and Pedersen discovered was that the quality of the recording was greatly improved if, during recording, a direct current was added to the recording head in a direction that served to reduce the magnetic state of the medium to less than saturation. This created a more balanced operating point about which to impose the recording signal. The use of dc bias in this way became standard in all magnetic recorders until the early 1940s, when it began to be replaced by the superior ac bias technique described in subsequent chapters.

Poulsen's assistant Hagemann was also involved in his own research on the possibility of using the Telegraphone as an amplifier. Hagemann's amplifier used a drum provided with a series of steel rings having their centers in the axis of the drum and their plane of orientation perpendicular to it. The signal to be amplified was recorded on the first ring as it rotated. The recording was then rerecorded onto the other rings in succession by copying from the first ring. The record on all the rings used to rerecord was the same strength, since the current in each recording electromagnet was the same. In principle, the device played back all the recordings simultaneously into the same circuit, producing a current identical to but stronger than the original input signal, provided the number of rings was sufficiently large. In practice, however, each additional signal added noise. Moreover, any slight misalignment of the playback coils meant that the components of the output signal would be out of phase with one another, adding more distortion. As a result, intelligibility quickly decreased to the point that the output, though louder, could no longer be understood. This problem was not immediately apparent, however, and when the Telegraphone was demonstrated in Paris, both dc bias and Hagemann's amplifier idea were central to the sales pitch of Aktieselskabet Telegrafonen and Mix & Genest.

The exhibition in Paris in the summer of 1900 was a major international industrial gathering. It provided the first opportunity for Aktieselskabet Telegrafonen and Mix & Genest to expose their Telegraphone to a large audience. At the exhibition, Aktieselskabet Telegrafonen constructed a small but elaborate kiosk in the section devoted to Danish industry. There the company demonstrated several machines of the cylinder type for the public (Fig. 3–3). Mix & Genest exhibited the improved Telegraphones in Paris at private demonstrations for invited guests, including reporters from the technical press. At these demonstrations, Poulsen and Pedersen were usually responsible for the operation of the equipment.

Pedersen and Poulsen's efforts during the demonstrations, as well as the work of Fog and Mix & Genest, paid off. The Telegraphone was the hit of the exhibition, receiving a great deal of publicity and a Grand Prix from the exhibits committee. A number of prominent individuals examined the machine, including members of the Siemens family and the novelist Emile Zola. Poulsen and Fog even managed to interest Emperor Franz Joseph in his machine; the Austro-Hungarian ruler consented to make a short recording. Presently preserved at the Danish Museum of Science and Technology, it is the oldest surviving magnetic recording in the world. Poulsen no doubt sought to emulate Alexander Graham Bell, who had induced the emperor of Brazil, Dom Pedro, to use his newly invented telephone 24 years earlier at the American Centennial Exposition in Philadelphia.

More importantly for the future success of Aktieselskebet Telegraphonen and Mix & Genest, a stream of technical articles appeared over the next few months in professional journals, describing the Telegraphone and singing its praises. All the articles stressed the many possible applications of the Telegraphone (as telephone answering machine, as a means for reporting news to tele-

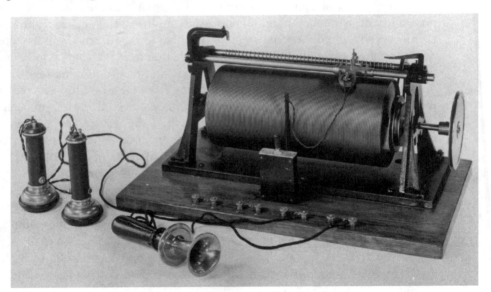

Figure 3–3   A Telegraphone similar to the machine demonstrated in Paris in 1900. The nonmagnetic cylinder was spirally wound with approximately 100 meters of fine steel wire. The recording head moved parallel to the cylinder axis on a threaded rod as the cylinder rotated, tracking the wire. Recording time was approximately 45 seconds. [*Courtesy of the Peder O. Pedersen Archive.*]

phone subscribers, as dictating machine), but it was the applications related to the telephone that received the most attention. Typical is an article on the news at the Paris Exhibition, by John Gavey, who later achieved the prestigious position of chief engineer of the British Post Office. The section on what Gavey called "Poulsen's Microphonograph" was introduced by the statement: "Perhaps the invention of the greatest scientific interest is the Poulsen Microphonograph, by which a telephone conversation can be permanently recorded on a steel wire, and reproduced at any time." A detailed description of the various functions of the apparatus was followed by the author's own judgment of its future prospects:

> At present this invention is in the early stage of scientific discovery. It may be used by a telephone subscriber to record an important communication, and it promises to afford means of obtaining a telephone repeater, a problem which has been before the electrical world for the last twelve years, and which so far has not been solved in a satisfactory manner. A telephone repeater would increase the range of telephonic speech and decrease the cost of long lines. The President of one of the American telephone companies some time ago offered publicly a reward of 1,000,000 dollars for a thoroughly satisfactory telephone repeater, but the money has not yet been earned.*

As is evident from this quotation, Hagemann's idea of using the Telegraphone as an electrical amplifier had caught the imagination of writers familiar with the telephone industry. No practical electronic amplifier existed at the time (the vacuum tube was still some years away), so any invention related to telephone amplification was hailed as a potential breakthrough. It appears that the Telegraphone amplifier concept, which was not actually demonstrated at the exhibition, accounted for a good deal of the extensive press coverage. Indeed, when subsequent tests showed that Hagemann's idea was unworkable as an amplifier, the interest of the technical journals in the Telegraphone declined markedly.

CONFLICT IN GERMANY

After returning from Paris, the researchers turned to developing a commercial machine. The primary focus was on improving the circuitry used to connect the Telegraphone to the telephone network. Despite some progress, the slow pace of development caused increasing tension between Mix & Genest and its Danish partners.

The conflict between the Danes and the Germans arose from the parties' divergent views about how to reach the goal of a commercial machine. Poulsen and

---

*Gavey (1900).

Pedersen felt that considerable basic research work remained to be done before the Telegraphone would be ready for service under field conditions. Managers at Mix & Genest, on the other hand, wanted a commercial product as soon as possible, and were unwilling to spend money on further work unrelated to producing a marketable machine. The collaboration was formally dissolved some time around the end of August 1900.

After the break with Mix & Genest, Lemvig Fog, the head of the Danish firm, began a search for new partners. In early 1901, he visited the United States, but was unable to raise any interest. Around the same time, Fog contacted a number of German firms, and in the spring of 1901 he reached an agreement with the prominent firm Siemens & Halske (ancestor of the current Siemens company). Siemens & Halske purchased the rights to produce Telegraphones in Germany, Austria-Hungary, and Russia. During the first few months of the agreement the Germans were very optimistic about the prospects for the Telegraphone, but by early 1902 their attitude had turned sour. Unable despite considerable investment to develop a lighter and simpler telephone answering machine with better sound quality, Siemens & Halske canceled their agreement in April 1902, claiming that the volume of reproduction was unlikely to become much better "unless a new way is found to fulfill this goal."

With the collapse of this second collaboration, Poulsen and his partner Pedersen abandoned work on magnetic recording. They turned their attention to radio, with considerable success. After inventing the arc transmitter for wireless telegraphy ("the Poulsen arc") in 1902, they spent the next 20 years developing it into a commercial product. Poulsen served as a member of the boards of the companies created to profit from his radio-related inventions, which were patented in most industrialized countries. He received many honors in his lifetime, including an honorary degree from the Technical University of Denmark, the institution whose entrance examination he had been unable to pass! Valdemar Poulsen died in Copenhagen in 1942.

A NEW DANISH STRATEGY

The unsuccessful ventures in Germany and the ongoing but seemingly endless negotiations with American investors led the Danish investors behind Aktieselskabet Telegraphonen to set up their own development and production company, Dansk Telegrafonfabrik, in 1903. The Danish Telegraphone factory over the next few years developed two different types of Telegraphone, a simple disk type with a recording time of approximately 2 minutes (Fig. 3–4), and a much more complicated wire type with a recording time of approximately 30 minutes (Fig. 3–5).

Both devices were designed as dictating machines, indicating a significant change in orientation among those behind the attempts to commercialize the

Figure 3–4 Danish disk-type Telegraphone, built in the period 1905–1909. It used a steel disk 0.5 mm thick and 130 mm in diameter, rotated by means of a clockwork drive. The velocity of the disk relative to the head was close to 0.5 m/s. [*Courtesy of the Danish Museum of Science and Technology.*]

Telegraphone. The Danish production company manufactured only a small number of these machines, probably fewer than 200.

The absence of a clear market strategy and absence of Poulsen and Pedersen, who had left the magnetic recording field to concentrate on Poulsen's new inventions within wireless telegraphy, account in part for this limited production. But the primary reason was that Dansk Telegrafonfabrik—which employed just a handful of people—was never meant to be a factory for large-scale production of Telegraphones. It was a precision workshop, set up to manufacture the delicate equipment needed by Poulsen and Pedersen in their increasingly successful work on wireless telegraphy, and—just as important—to improve the bargaining position of the Danes in the ongoing negotiations with American investors.

NEGOTIATIONS WITH AMERICAN INVESTORS

The first Danish delegation to the United States actually had been sent in January of 1901. Publicity from the Paris exhibition and the contacts gave the delegation ready access to interested American investors, including representatives of the

Figure 3–5   Danish wire-type Telegraphone, built in the period 1905–1909. It used 0.25 mm diameter steel piano wire, moving with a velocity of 2 to 3 m/s with respect to the head. [*Courtesy of the Danish Museum of Science and Technology.*]

Bell Telephone Company and men who had financed Edison's inventions. The delegation spent most of their time over the next few months meeting with potential investors and demonstrating the machines.

After long and difficult series of negotiations, an agreement was reached in March 1903. The agreement formed the basis for the incorporation of the Telegraphone Company of Maine in September 1903. This company in turn granted the American patent rights to an American-owned firm, the American Telegraphone Company. However, the American investors who established this firm proved unable to raise sufficient money to set up a factory, and no development work was done for almost two years.

THE AMERICAN TELEGRAPHONE COMPANY

In the summer of 1905 Charles Fankhauser, an American broker and investor, persuaded the Danes to transfer a majority holding of their stock from the Maine corporation to the American Telegraphone Company's own treasury in exchange for a small cash payment. This gave the board of directors of American Telegraphone direct control over their company, since the Danes were now effectively minority shareholders. This ended direct Danish involvement in the Telegraphone's development. Fankhauser had bought six recorders from the Danish company. Using these machines as demonstrators, Fankhauser was able to sell much of the stock to small investors, raising over $300,000 for American Telegraphone (and apparently making at least that much or more for himself through high commissions). This sum, the equivalent of about $5 million today, allowed American Telegraphone to set up a factory and begin to develop its own products.

The company decided to focus on dictating machines. Since secretaries could use earphones during the transcription of recorded messages, the low volume of reproduction was a less serious problem. Moreover, the rising interest in scientific management in the United States had led to the increasing use of dictating equipment by large corporations at this time. Throughout its years of operations, the primary focus of American Telegraphone's sales efforts was the dictating machine market.

Poulsen and his associates had produced two types of Telegraphone at Dansk Telegrafonfabrik in Copenhagen. A reel-to-reel machine using piano wire as a recording medium had the advantage of a long recording time. By 1905, the prototype could record for up to 30 minutes. The wire was also cheap and easy to procure. If the machine was not kept in exact adjustment, however, it would break or snarl its wire. The Danish wire machines were also extremely complicated mechanically; the ones available in 1905 contained over a thousand parts. Obviously, this made both manufacture and maintenance complicated.

The second type of machine, a disk recorder, had advantages and disadvantages that mirrored those of the wire machine. The disk recorder was mechanically much simpler (it used a windup clockwork mechanism to rotate a thin steel disk), and there were no problems with breakage or snarling. However, the disk machine did have some notable disadvantages. The recording time for the 5-inch-diameter disk was only 45 seconds per side. Also, unlike wire, flat steel disks were not a common commodity and so would have to be specially manufactured, placing an additional burden on the company.

After considering the advantages and disadvantages of both types of machine, the board of directors mistakenly decided to focus on the disk version. They felt that the simplicity of the disk machine would enable the company to put a product on the market much faster, and so earn a quicker return. To increase recording time, they planned simply to scale up the diameter of the disk from 5 inches to 12 inches, but this proved to be more difficult than anticipated. Interference from the motor then required a major redesign to incorporate shielding. Finally, the new larger disk tended to warp during heat treatment, and thus was more difficult to manufacture.

These problems were eventually solved in a technical sense, but the resulting machine was essentially unmarketable. It had a recording time of only 2 minutes per side, 4 minutes total per disk, and would have sold for $300. A contemporary phonograph, costing only $70, could record for about the same length of time. Not more than one hundred disk recorders were ever built, and there is no evidence any of them entered commercial service.

By May 1908 the company had run out of money, and Fankhauser helped the board of directors to arrange for a friendly takeover. The new investor was Edwin Rood, former president of the Hamilton Watch Company. Rood purchased a large block of American Telegraphone stock and was installed first as president and soon after as general manager. Within two years, Rood had taken complete control of the firm and had replaced all the members of the board with his own associates.

Rood's first move was to abandon work on the disk recorder and shift the firm's efforts to producing a wire recorder. As well as ordering the completion of a prototype based on that work, Rood bought the rights to another machine developed by the inventor George S. Tiffany and financed by Fankhauser.

Over the next three years, both American Telegraphone and Tiffany developed machines independently. In February 1911, Rood comparison-tested his firm's latest design and a new machine developed by Tiffany. Although American Telegraphone's recorder had more features, the other machine had superior sound quality and was much cheaper to manufacture. Rood purchased the rights to manufacture the Tiffany machine, and preparations were made to start production of this design (Fig. 3–6). The first production machines left the factory in August 1912.

The machines were marketed by the Telegraphone Sales Company, an independent marketing firm set up the year before and founded by the McCrillis family, who entered into a contract with Rood to buy American Telegraphone's entire production and to set up a series of sales offices in major cities.

McCrillis was initially successful in selling around one hundred Telegraphones. The single largest sale was to the Du Pont Company, which bought 20 machines initially and almost 30 more over the next two years. This installation, opened in 1913, included a telephone network, which allowed executives in their offices to call up a central switchboard and record dictation for later transcription. Although initially happy with the system, Du Pont stopped using it in 1917 and scrapped it in 1919, largely because of continual breakdowns and the difficulty in getting replacement parts. Similar complaints dogged the other machines McCrillis sold, and he was out of business by 1915.

There is no question that the weakness of the reproduced audio signal from a Telegraphone was a major factor in the lack of commercial success. It is interesting to note, however, that electronic amplification of a Telegraphone signal was carried out in 1913 by de Forest, the inventor of the "Audion" amplifier. He published the results in 1914 and claimed that a violin record on a Telegraphone could be heard at a distance of 259 feet in quiet open air with the help of his amplifier. But de Forest did not persevere with his experiments, and perhaps they came too late to revive the already failing fortunes of the Telegraphone.

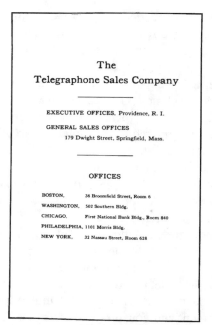
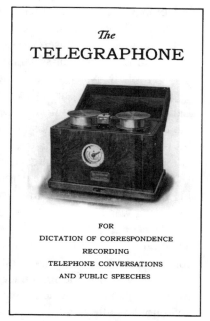

The
Telegraphone Sales Company

EXECUTIVE OFFICES, Providence, R. I.

GENERAL SALES OFFICES
179 Dwight Street, Springfield, Mass.

OFFICES

BOSTON,   36 Broomfield Street, Room 6
WASHINGTON,   502 Southern Bldg.
CHICAGO,   First National Bank Bldg., Room 840
PHILADELPHIA, 1101 Morris Bldg.
NEW YORK,   32 Nassau Street, Room 628

The
TELEGRAPHONE

FOR
DICTATION OF CORRESPONDENCE
RECORDING
TELEPHONE CONVERSATIONS
AND PUBLIC SPEECHES

Figure 3–6   The Model 900 American Telegraphone recorder, the final version produced by the ill-fated company. Maximum recording time was 30 minutes, with a wire speed of 3 m/s. Recording was done through a telephone handset (not shown); the dial on the front of the device indicated total recording time. During playback, a typist wearing a headset used a foot control to deliver three commands: forward, stop, and reverse. [*Courtesy of the Peder O. Pedersen Archive.*]

By the late 1910s, the competing technology of the wax cylinder dictating machine was much cheaper and more reliable than American Telegraphone's machine, whose advantages of longer recording time and easy telephone hookup for centralized dictation did not compensate for its basic unreliability. American Telegraphone's competitors, the Dictaphone and Ediphone companies, paid close attention to both manufacturing and marketing their phonograph dictating equipment, changing their machines to make them easier to use and more reliable. American Telegraphone would have had to match that performance if it wanted to compete. Rood's management style proved unequal to the challenge.

By 1917 the company's only customer was the United States Navy, which used Telegraphones for recording high-speed wireless telegraphy signals. The navy was unaware of this potential application until 1916, when it was discovered that a German-owned radio station in Paterson, New Jersey, was using a speeded-up Telegraphone to send intelligence information to German submarines operating in the Atlantic. When German spies recorded on a Telegraphone at one speed and then speeded the recording up for transmission, the signal became an indecipherable whine, which could be decoded only by recording at high speed at the receiving end and then slowing down the recording. The Germans used Telegraphones purchased prior to the first world war from American Telegraphone through Poulsen's Danish company.

The U.S. Navy set up a series of experiments on recording wireless telegraph signals after purchasing the machines. Carried out at the Naval Research Laboratory in Washington, D.C., this research led to the discovery of the principle of ac bias, the superposition of a high-frequency signal on top of the signal to be recorded, resulting in improved quality. The inventors, W. L. Carlson and Glenn L. Carpenter, applied for a patent in 1920, and it was granted in 1927 as U.S. Patent 1,640,881. The patent covered the application of ac bias to the recording of telegraph signals only, not sound. As a result, ac bias for voice and music recording was later discovered and patented separately.

Despite the failure to sell more than a handful of Telegraphones, American Telegraphone survived by becoming a subcontractor for precision machine work for the military during the first world war. Even with war work, however, the company was floundering under a large debt load. The company was trying to sell off its assets in 1918 when it was sued in the District of Columbia by a number of shareholders alleging mismanagement and conspiracy to defraud. The court froze the company's assets and appointed a receiver to manage the company.

At first the company continued to sell Telegraphones (about 40 machines were sold between 1920 and 1924), mostly to large corporations whose interest in magnetic recording was piqued by the pending expiration of Poulsen's basic patents, but eventually that income dried up. The court case dragged on and on, and in 1933 all the company's patents ran out. After that, the company ceased to control any marketable assets and was inactive until the formality of its court-ordered dissolution in 1944.

CONCLUSION

Valdemar Poulsen and his fellow Danes foresaw a brilliant future for the Telegraphone as a new and potent component in the rapidly growing telephone system. But as the road to the promised land seemed blocked by almost unsurmountable difficulties, they soon gave up and left it to the Americans to develop a "standard Telegraphone." Edwin Rood saw the Telegraphone as a new and advanced dictating machine, able to compete effectively with the phonograph in the business administration market. Rood's strategy was basically sound, but his unrecognized problem was the belief that the Telegraphone was a machine to be produced and marketed like any other standard commodity, rather than a new technology to be nurtured. Trapped within their inadequate visions of this new technology, neither group was successful.

REFERENCES

Clark, Mark, and Henry Nielsen, "Crossed Wires and Missing Connections: Valdemar Poulsen, The American Telegraphone Company, and the Failure to Commercialize Magnetic Recording," *Business History Review,* **69**, 1–41 (1995).

Gavey, J., "Poulsen Microphonograph," *Journal of the Institution of Electrical Engineers*, **30**, 88–89 (1900).

Larsen, Absalon, "Telegrafonen og den Traadlose," *Igeniorvidenskabelige Skrifter,* No. 2, 1950.

Poulsen, Valdemar, "Method of Recording and Reproducing Sounds or Signals," Danish Patent 1260 (Dec. 1, 1898).

Poulsen, Valdemar, "The Telegraphone: A Magnetic Speech Recorder," *The Electrician,* **46**, 208 (1900).

4 Steel Tape and Wire Recorders

Mark H. Clark

Between 1920 and 1945, the first commercially successful magnetic recorders were designed and sold. With the exception of the German Magnetophon, described in the next chapter, these machines used solid steel recording media, either in the form of tape or wire. They were designed and built in three countries: Germany, the United Kingdom, and the United States. In addition, research was carried out in Sweden and Japan.

In Germany steel tape and wire recorders were used for both business and broadcast purposes. In the United Kingdom, steel tape recorders were developed for radio broadcasting and sold around the world. In the United States, wire recorders were sold in large numbers to the American military during the second world war and, immediately after the war, to civilians for voice and music recording. This chapter describes the development of steel tape and wire recorders after 1920, and the subsequent impact of these devices on magnetic recording technology.

GERMANY

Between 1920 and 1945, Germany was the most active site for magnetic recording research in the world. The man initially responsible was Kurt Stille, a German inventor and entrepreneur. He set up the first commercially successful companies that produced magnetic recorders, and his work led several large electrical firms to enter the market, either by buying Stille out or by developing competing products. As a result, by the 1930s magnetic recording was firmly established commercially in Germany, and German firms marketed dictation machines and studio recorders intended for radio use domestically and abroad.

Stille first encountered magnetic recording in 1903, when he bought a Telegraphone from Mix & Genest, the German firm that had worked with Valdemar Poulsen to develop his invention. Stille, an electrical engineer, was attempting to develop an electronic amplifier, and he used the Telegraphone in conjunction with his experiments. These proved unsuccessful, and he set aside the Telegraphone in favor of other work. Over the next several years he patented a variety of electrical devices, including a relay design, a photocell system, and a facsimile machine.

After the first world war, Stille returned to magnetic recording, apparently motivated by the pending expiration of Poulsen's basic patents. In 1920 he tried to market a modified American Telegraphone wire recorder in Germany, but without success. During the early 1920s, he obtained a number of patents on his designs for a magnetic recorder and formed several patent holding companies to exploit them. One of these companies, Vox Maschinen-Aktiengesellschaft, created in 1924, was a sales organization for dictating equipment. The following year, Stille developed an improved wire recorder intended for Vox Maschinen-AG. Much like the American Telegraphone machine, it used piano wire, running at approximately 2 m/s between two large spools. This new machine attracted considerable attention in the technical press. The recorder incorporated a number of innovations, most notably an electronic amplifier that made listening to recordings much easier. The wire spools were enclosed in a metal shell, allowing for easy replacement (much like the present-day cassette recorder). In addition, provisions for telephone connection allowed the machine to serve as a central station dictating device or as an answering machine.

Unfortunately for Stille, the economic climate of late 1920s Germany was not favorable for the introduction of his machine. Despite good reviews in the business press, it proved too expensive, especially in a period of high unemployment, which meant that the wages of stenographers were low and declining. As a result, sales were small.

However, the attention he received did create other opportunities. In 1929 he attracted the interest of Ludwig Blattner, a German film producer working in England. Blattner wanted to use magnetic recording for a motion picture sound system, and he purchased the rights to Stille's patents for a substantial sum. The resulting machine, the Blattnerphone, is described in more detail later.

Stille continued to work on improving the dictating machine. In 1928 he entered into a partnership with one of the men who had licensed his patents, Karl Bauer. Through the newly incorporated Echophone company, which was intended as a marketing and sales organization, Stille and Bauer contracted out the actual development and manufacture of the machine to Ferdinand Schuchard AG, a telephone equipment manufacturing firm. Schuchard, which lacked a research department, hired Semi Begun, an electrical engineer, to design the electronic circuitry. Begun soon became the firm's specialist in magnetic recording.

Stille had contracted for a machine whose performance would be superior to contemporary wax cylinder dictating machines. This meant a recorder with reasonably flat response from 250 to 3000 Hz and a signal-to-noise ratio of 30 dB or more. In 1930, Begun and his coworkers produced a prototype that met the standard, and the Echophone board decided to place it into production. They named the machine the Dailygraph, and set out to market it to businesses in Europe.

Begun left Schuchard and went to work directly for Echophone as a development engineer, with responsibilities for meeting with clients and incorporating their suggested improvements into the machine where possible. He traveled widely in Europe over the next two years, though since the machine was fairly expensive there were few sales. As Begun later recalled, most of those who purchased a Dailygraph were attracted by its novelty and high-tech image rather than by its practical advantages.

In August 1932, Stille and his partner Bauer sold the Echophone company to International Telephone and Telegraph. ITT bought Ferdinand Schuchard AG at the same time and incorporated both firms into its German subsidiary, C. Lorenz AG. Stille, well off due to the sale and other investments, retired and did no further work in magnetic recording.

Begun went to work for Lorenz, and was placed in charge of a development program in magnetic recording. His first task was to redesign and upgrade the Dailygraph, incorporating improvements to correct problems he had encountered while repairing machines in the field. The new machine was released in 1933 as the Textophone (Fig. 4–1). Similar in design and operation to the Dailygraph, the

Figure 4–1   Textophone built by the Lorenz company in the mid-1930s. The recording wire, mounted on two large reels, is contained in the cassette at the top of the machine which can be removed as a unit. [*Courtesy of the National Museum of American History, Smithsonian Institution.*]

Textophone was intended for dictation and recording telephone conversations. Several thousand machines were produced, the single largest customer being the German government. Lorenz continued to sell the Textophone, essentially unchanged in design, until 1939.

Lorenz also produced a recorder for radio broadcast use. Called the Stahltone-Bandmaschine, it was first offered for sale in 1935. It used a steel tape, running at 2 m/s, giving a reasonable frequency response between 70 and 5500 Hz. This performance was good enough to allow voice recording for AM radio broadcast. The machine was intended for mobile recording, but because of its considerable weight (150 kg), it was normally truck-mounted or installed in temporary recording studios.

After a series of tests, the German state broadcast service, the Reichs-Rundfunkgesellschaft (RRG), purchased several of the Lorenz machines. The same year, Lorenz provided a Stahltone-Bandmaschine and six Textophones for press use at the 1936 Winter Olympic games. Lorenz also sold recorders to the radio station run by the Bureau des Postes, Télégraphs, et Téléphones in Paris, and to the Swiss state radio service.

Begun wrote an account of his magnetic recording work in 1933, earning a doctorate in electrical engineering from Berlin's Technische Hochshule. Although he was well regarded by his superiors at Lorenz, Begun's liberal political views and Jewish background made him feel increasingly uncomfortable in Hitler's Germany, and he emigrated to the United States in 1935.

Lorenz continued to sell recorders after Begun's departure, but no new research projects were undertaken. Increasing competition from the Magnetophon, which used coated plastic tape, gave the German government an alternate source for recording equipment. Both the Textophone and the Stahltone-Bandmaschine were withdrawn from the market in 1939.

THE UNITED KINGDOM

The initial focus of research in magnetic recording in England was to develop a system for adding sound to motion pictures. During the 1920s, a variety of movie sound systems competed in the marketplace, such as phonographic and sound-on-film technologies. The Blattnerphone, based on technology developed by Stille and his associates, was originally intended for the same market. Although never used commercially for motion picture sound, the Blattnerphone eventually formed the basis for recorders used during the 1930s and 1940s by the British Broadcasting Corporation and other broadcasters.

As noted earlier, Stille had had difficulty selling his recorders for office dictation during the period of hyperinflation in Germany in the 1920s. As a result, in 1928 he began to develop a motion picture sound system using his knowledge of magnetic recording. Stille was not the first to attempt to use magnetic recording for movie sound. Lee de Forest had suggested the possibility in 1913, as a result of his work with the Telegraphone. Between 1915 and 1928, five other Americans developed similar systems, but none advanced beyond the prototype stage.

Stille was more successful, largely because he was able to demonstrate his dictation machines to potential investors like his countryman Ludwig Blattner, who bought the rights to use the new sound system in 1929. A former movie exhibitor, who had emigrated to England, Blattner had recently entered the movie production business, hoping to take advantage of a change in British law that favored domestically made films. Blattner wanted to combine Stille's machine with another new technology, the recently developed Keller–Dorian lenticular color process, to make his films more attractive to audiences. Blattner renamed Stille's machine the Blattnerphone and set out to interest film exhibitors in its use.

The machine Blattner purchased was substantially different from the dictating machines Stille had produced previously. Instead of wire as a recording medium, it used steel tape. The tape, which was 6 mm wide and approximately 0.05 mm thick, ran at roughly 2 m/s. Its most distinctive feature was a row of perforations down its centerline, which meshed with sprockets on the machine's drive capstan. The sprockets seem to have been included more to give the appearance of synchronization than to aid its actual accomplishment. In practice, to keep the tape in sync with the film, an operator had to adjust a rheostat constantly during projection.

The need for constant monitoring, combined with Blattner's lack of capital compared to his competitors in the sound film market, prevented the Blattnerphone from ever reaching the market. Blattner was reduced to performing novelty acts in music halls with his machine, recording the impromptu performances of members of the audience and playing back the results.

Fortunately for Blattner, he was able to find another market for his machines. In 1930 the British Broadcasting Company purchased a two-year option on the rights to use the Blattnerphone and in 1931 began to rent machines from Blattner for broadcast use. The BBC had been interested in magnetic recording since the mid-1920s, as a solution to problems associated with its new shortwave broadcasting program, the Empire Service. Broadcasting at the best time for reception in the various target countries (typically early evening, local time) would have required a large staff working around the clock, even though essentially identical material was sent to each country. Recording and repeatedly replaying material during the broadcast day was seen as a way to reduce staffing costs.

Although recording could be done with phonographic disk machines, this option was not without a number of limitations, most notably the need for a highly skilled operator and the limited recording time of only a few minutes. Sound on film offered longer recording times, but the photographic film available had to be chemically developed before use. In contrast, the Blattnerphone offered immediate playback, long recording times, and less difficult operation. Thus, the planned 1932 start of the Empire Service inspired the BBC to pursue extensive testing of the Blattnerphone for broadcast use.

The machines installed in the BBC's Savoy Hill studio in 1931 were essentially identical to the ones intended for sound film use, except that they omitted the perforations in the tape. Despite the speed stability problem, the BBC found the machines satisfactory, and plans were made to use them when the Empire Service went on the air. The following year, BBC engineers and Stille Inventions Ltd. improved the transport system by replacing the dc motor with a synchronous ac motor, eliminating the need for operator speed monitoring. They also improved the equalization circuitry and reduced the tape width to 3 mm. In this modified form, the Blattnerphone was used by the BBC until 1936.

After prolonged litigation, Blattner's motion picture production company had gone into receivership and was finally wound up in 1933. As part of the liquidation, Blattner sold the rights to the Blattnerphone to Marconi's Wireless Telegraph Company, Ltd. A broken man, Blattner committed suicide two years later.

Marconi, in cooperation with Stille, developed a second-generation machine for sale to radio broadcasters in other countries. The new machine was essentially identical to the Blattnerphone as modified by the BBC, with the addition of backup recording and reproducing heads. The heads each used two pole pieces, one on each side of the tape, rather than today's common ring heads. Erasing and recording heads were made from Stalloy, the reproducing heads from Permalloy. To confine the magnetic flux generated on the tape, each pole piece was sharply tapered at the point of tape contact. The tape wore the pole pieces down rapidly, which led to the inclusion of the backup heads. The tape was made from high-quality Swedish steel, processed to minimize variations in structure, which had caused problems with the Blattnerphone's performance. The Marconi–Stille machine was sold to radio stations in Canada, Australia, Egypt, Poland, and France.

In 1935 Marconi, in collaboration with the BBC, developed a third-generation recorder (Fig. 4–2). Dual-pole recording heads were still used, but the reproducing head was changed from dual-pole to single-pole design. The transport mechanism was improved by incorporating three separate drive motors, and the electronic circuitry was upgraded. Such machines were sold to the BBC and to Radio-tjanst in Sweden. The performance of the Marconi–Stille machines was excellent for the time: a response flat to about 6 kHz with a signal-to-noise ratio of 35 to 45 dB.

The recorders installed at the BBC were used for a variety of tasks. Their primary role was that for which they had been designed—the recording and reproduction of Empire Service shortwave broadcasts. They were also used for recording material for national rebroadcast, particularly important speeches by politicians or the king. Tapes of important programs were often archived for future use. Finally, announcers used the recorders during training, and for rehearsals before important performances.

The recorders installed at radio stations outside England were used in a similar fashion, though with some differences. Recorders in Canada and

Figure 4–2   Marconi–Stille recorders of the type used by the BBC. Pairs of recording and reproducing heads are mounted in a horizontal array at top center. To the left of the heads is a small tape reservoir, and a large tape reservoir is installed below. This arrangement served to isolate the tape in the head region from the effect of the "unwinding and winding systems." [*Courtesy of Ampex Archives.*]

Australia were used to record Empire Service shortwave programs for later retransmission on domestic radio service. The French station used the machines to record and repeat programs for domestic use. The Polish and Egyptian radio systems made relatively little use of their recorders—apparently they were purchased as part of a package deal by governments whose broadcast facilities had been built by Marconi.

Neither the Blattnerphone nor the Marconi–Stille machines were easy to use. Simply mounting a reel of tape was a two-person operation, since a full reel contained 2700 meters of tape and weighed over 35 kg (Fig. 4–3). In operation,

Figure 4–3   A rack of Marconi–Stille reels was an impressive sight. Each full
reel contained 2700 meters (about 1.7 miles) of steel tape 3 mm
wide and 0.8 mm thick. [*Courtesy of Ampex Archives.*]

the rapidly moving tape was much like a band saw, with the potential for slicing
off stray fingers. It was possible to edit material by cutting tape and silver-solder-
ing sections together, and this was a common practice by the mid-1930s. Later
models had a built-in spot welder. The tape, being of hardened steel, was brittle,
and breakage often occurred, spilling tape into the room. Despite the hazards, all
these installations continued in use through the late 1940s, when they were replaced by other machines using coated rather than solid steel tape.

THE UNITED STATES

As a result of the legacy of the American Telegraphone Company, the United
States continued to be an active area for magnetic recording developments. When
Poulsen's patents expired in the 1920s, numerous firms and individuals began to
experiment with magnetic recording. Most, though not all, of these efforts were
unsuccessful.

Three American companies were able to build on the work of Poulsen and
produce magnetic recording products prior to the second world war: American
Telephone and Telegraph (AT&T), the Brush Development Company, and the
Armour Research Foundation. AT&T built machines for its own use within the
telephone system, taking advantage of the expertise of its researchers at Bell

Laboratories. The Brush Development Company became involved in magnetic recording after it hired Semi Begun, the former director of the Lorenz magnetic recorder program in Germany. The Armour Research Foundation built a patent licensing system based largely on the work of the engineer Marvin Camras.

Bell Telephone Laboratories

Engineers at AT&T were among those interested in magnetic recording during the 1920s. However, it was not until 1930 that Bell Telephone Laboratories (Bell Labs), the AT&T research arm (now Lucent Technologies) initiated a project to investigate the application of magnetic recording to various areas, including sound delay, tone generation, and prerecorded announcements. Bell Labs director F. B. Jewett hired Clarence Hickman, a physicist, and assigned him full time to magnetic recording research.

Hickman began by experimenting with an American Telegraphone wire recorder the laboratory owned. He tried several ideas for improving quality, such as varying wire size and altering the design of the head. Within a month, however, he abandoned work on wire and switched to steel tape as the recording medium. He was convinced that unavoidable twisting of the wire between recording and playback was responsible for considerable degradation in performance. Tape, which always presented the same geometry to the heads, would not suffer from this problem.

Hickman's research was along the same lines as that conducted in Germany and England. In cooperation with others in the laboratory, he developed new steel alloys and heat treatment methods designed to improve the magnetic properties of steel tape for recording purposes. He also experimented with various arrangements for the recording and playback heads. While no particular innovation stands out, a series of small improvements resulted by late 1930 in a recorder with good sound quality.

Over the next several years Bell Labs moved to begin building laboratory prototypes of machines for different purposes: a telephone message recorder, a portable tape recorder, a sound delay recorder, an endless loop recorder, and several others. Based on the experience gained with these prototypes, Bell Labs built several machines for field testing between 1935 and 1940. These included a telephone recorder for news agencies and a telephone announcement system used by the New Jersey State Department of Agriculture. The latter machine was further developed and widely used for automatic announcement in the Bell System, most notably for a weather forecast announcing system operated by the New York Telephone Company from 1939 onward.

Given this level of success, why did AT&T fail to market its recorders in competition with the Dailygraph, Blattnerphone, or Stahltone-Bandmaschine? There seem to be at least three reasons. First, since AT&T already sold high-quality phonographic disk recorders through a subsidiary, sales of magnetic recorders to radio stations or sound studios would have cannibalized the market for those products. Second, the major American radio networks favored live over prerecorded programming, and so were not a good market for high-quality recorders. Third, AT&T managers were uncomfortable with recorders outside their own control. They were convinced that their customers would use the telephone less often if they knew that their conversations could be recorded—one executive was quoted as saying that one-third of all telephone conversations were either immoral or illegal, and that was too much business to risk losing.

The focus of AT&T's interest in magnetic recording changed with the coming of the second world war. Work at Bell Labs was shifted from serving the telephone industry to war work, and magnetic recording was caught up in that same change. Between 1940 and 1945, Bell Labs and Western Electric (the manufacturing arm of AT&T) designed and built a variety of recorders for the American military.

The most widely used Bell Labs device using magnetic recording was the "Heater" series of acoustic deception systems. The first in the series was developed for use in naval landings. Contained in a standard torpedo tube, the device was designed to surface, expose a loudspeaker, and play the sounds of battle to the enemy. Deployed in areas away from the main landing, the device was intended to confuse the enemy as to the exact place of landing and so slow the movement of reinforcements. During development, the device came to be known as the "Water Heater" because of its resemblance to a civilian water heater. Bell Labs also developed a series of similar devices for use on land and to broadcast propaganda from aircraft.

Despite its wartime work with magnetic recording, Bell Laboratories and AT&T in general played a limited role in postwar developments. As magnetic recording shifted to coated tape, away from the solid tape familiar to Bell engineers, AT&T came to rely more and more on outside suppliers for its magnetic recording equipment. For example, when AT&T finally offered magnetic recording telephone answering machines to its customers in the early 1950s, the equipment was made by outside contractors.

The Brush Development Company

In the 1930s, the Brush Development Company of Cleveland, Ohio, was a major manufacturer of piezoelectric crystals and supplied most of the crystal phonograph pickups used by the American record player industry. In 1938 Semi Begun was hired to head the company's engineering department, with the primary objective of expanding the Brush product line.

Over the next three years, Begun introduced Brush to magnetic recording and, in early 1939, the first commercial product was produced. Named the

Soundmirror, the machine used an endless loop of steel tape driven at a speed of 1.1 m/s, giving a total recording time of 50 seconds. Except for a microphone, the components, including a loudspeaker, were mounted inside a single case.

Although sales of the Soundmirror were limited, Brush's directors were sufficiently encouraged to expand the company's research program. Over the next two years, Begun and his assistants developed an electrical transient recorder (for General Electric), an artificial reverberation device, a speech scrambler, and a slow-moving tape drive. By the end of 1941, progress had been made on all these projects, but none of them were ready for commercial production.

America's entry into the second world war changed the nature of work at Brush. The outbreak of hostilities brought in a flood of orders, at first for phonograph crystals, and then for radio-related equipment. By the end of 1942, about 75% of Brush's sales were to the U.S. government, primarily to the military. By 1945, yearly sales of recorders and component parts, a negligible sum in 1942, had risen to almost $2.8 million. Brush built recorders for all three military services, and by the end of the war the firm was acting as subcontractor for all the recorders produced by Western Electric. According to the company's annual report for 1945, Brush produced the "vast majority" of all magnetic recorders purchased by the government during the war, including wire recorders designed for use aboard reconnaissance aircraft, and small portable recorders for use by Army scouts.

One government contract for research and development work on coating nonmetallic bases with magnetic materials led to several important innovations with postwar consequences. Initial activity focused on electroplating. By the end of 1943, Brush Development had successfully electroplated a cobalt–iron alloy onto a steel base, resulting in a 20 dB improvement in the signal-to-noise ratio. Later, Begun obtained funding from the Army Air Force for a program to reduce the weight of airborne recorders by improving recording materials. Work focused on producing a plated wire, and a product with commercial potential eventually emerged. Brush marketed electroplated wire for recording purposes after the war. Brush engineers also pursued work on vacuum deposition and suspended particle coating, though almost all that work was contracted out. Vacuum deposition proved unworkable with the technology of the time, but particle coating worked out well and came to serve as the basis for most of Brush Development's postwar products.

With the end of the war, Brush focused mainly on developing particle tape based recorders. However, the company did manufacture wire recorders for civilian use. The most significant was a specialized recorder capable of playing for 3 hours, intended for use on railroad cars or in industrial plants to provide "canned" music. Roughly a thousand machines were sold before production ended in 1950.

The Armour Research Foundation

The Armour Research Foundation in Chicago was the third center for magnetic recording research on wire recorders. Marvin Camras, an engineer and inventor at Armour, developed a number of magnetic recording innovations. Armour licensed Camras's patents to an expanding group of companies during and after the second world war. The licensee network that Armour built was important in the dissemination of magnetic recording technology in the United States.

Armour was set up in 1936 by administrators at the Armour Institute of Technology (AIT), a technical college in Chicago. Armour Institute, which became part of the Illinois Institute of Technology in 1940, wanted to foster relationships between its faculty and corporations in the Chicago area. Formed as a nonprofit institution, Armour was designed to provide a mechanism whereby AIT's faculty and staff could carry out research for businesses.

Marvin Camras started work at Armour in 1940, after receiving a degree in electrical engineering from AIT. Camras later attributed his initial interest in magnetic recording to a cousin who often sang in the bath. Believing that his voice was of operatic quality, the cousin asked Camras to build him a recorder for voice practice. Unable to afford steel tape, Camras built a wire recorder, but with one significant innovation. Camras realized that a wire subjected to a more uniform magnetic field around its circumference would reduce the distortion due to twisting of the wire between recording and playback, and he sketched out a design for a recording head that would create a more symmetrical field.

Camras finished the prototype in the spring of 1939. The machine worked extremely well, even though it was only a breadboard model. Camras's cousin abandoned his opera singing career soon after hearing his voice played back, so the machine caused some disappointment. The outcome for Camras, however, was much more favorable: the Armour Research Foundation offered him a job and the chance to develop his ideas about magnetic recording. Camras's primary responsibility after he was hired at Armour was research and development. Camras's work was financed initially by Foundation funds in anticipation of profits from manufacturing. Eventually, income from Camras's patents paid for the research program.

During the summer of 1940, while experimenting with recording high frequencies, Camras noticed that his recorder suddenly became more sensitive when he used frequencies above 15 kHz. He had rediscovered ac bias. Realizing immediately that he had made a major discovery, he moved to refine the technique and to incorporate it into the designs of recorders being planned for production. Camras applied for a patent on his innovation in December 1941. Although the Camras patent was clearly anticipated by work in Japan and paralleled by work in

Germany, wartime limitations on patent searches meant that the Armour inventor got his patent anyway.

Regardless of the patent's legitimacy, the invention had considerable practical effect. The combination of improved heads, better wire, and ac bias made for a lighter, more compact magnetic recorder with greatly improved fidelity. When the United States entered the second world war, Armour was in an excellent position to capitalize on Camras's inventions and supply the new military market. Camras and his colleagues had started work on their first commercial recorder, the Model 50, in late 1940 (Fig. 4–4). The civilian prototypes served as the basis for a militarized version of the Model 50.

Armour set up a small manufacturing facility at the institute's research site. Originally intended to be a stopgap measure, this production facility continued to operate until the end of 1944, when General Electric took over production at its own facilities. Armour built approximately 2000 Model 50 recorders for the military. Exact production figures for General Electric are not available, but the company built approximately 6000 of the Model 51, a copy of the Model 50.

These wire recorders served in a wide variety of ways, from the training of antisubmarine warfare crews to use as dictating machines, and for recording military news reports. Armour also loaned recorders to two national radio networks,

Figure 4–4   Model 50 wire recorder, designed by Marvin Camras and made by Utah Radio. [*Courtesy* of *Ampex Archives.*]

NBC and CBS, for use at the 1944 Democratic and Republican conventions, although it is unknown whether recordings made on those machines were aired. The Model 50 and its offshoots played a significant role in establishing Armour's postwar position in the magnetic recording industry. One model had an extended high-frequency response for recording music. That machine's enhanced performance, and its early association with General Electric, led to Armour's licensing more and more firms as the war progressed.

By late 1944, pressure to produce recorders for the military had eased. Armour committed itself to developing equipment for commercial purposes. Although some work on producing prototypes continued, the company's major function became increasingly that of a central research laboratory for its licensees. As such, Armour focused on areas where it had particular expertise, namely, the design of recording and playback heads and the development of improved recording media.

By the end of the second world war, Camras had developed the system that was to form the basis for most consumer wire recorders built after 1945 in the United States. The design made use of commonly available components to keep manufacturing costs low. The recorders proved to be popular, and more than a dozen firms produced wire recorders based on the Camras design between 1946 and the early 1950s. Annual sales were in the tens of thousands of units, with recorders generally selling in the $100–$200 range. Webster-Chicago was one of the more successful firms, with a line of portable equipment.

Wire recorders were largely replaced by tape recorders in the marketplace by 1950, and wire recorder makers either went out of business or switched to tape recorders. Nevertheless, by 1950, Armour had 53 licensees, including virtually all the companies making magnetic recorders in the United States. Armour continued to obtain more licensees throughout the 1950s, including Japanese companies. This licensee network allowed the dissemination of information about magnetic recorder manufacture and also made Armour a great deal of money. By pursuing a licensee strategy, Armour was able to exploit its patents despite its initially weak financial position, establishing a leading role in research that would continue into the 1960s.

SWEDEN

Sweden was the primary center for research into the manufacture of steel recording tape in the 1930s. The German inventor Kurt Stille first approached the Swedish steel company Uddeholm AB in 1930 about the manufacture of recording tape for the Blattnerphone, the magnetic recorder used by the BBC. Tests had shown that variation in reproduction volume was caused mainly by variations in tape quality, both in terms of chemical composition and dimension. Uddeholm was well known as a supplier of specialty steel, and when an experimental batch it produced for Stille proved successful, Stille asked the firm to enter into volume production.

After additional research by Uddeholm, the firm commenced manufacture in 1931 of a special grade of steel suitable for use as magnetic recording media. Referred to as type UHB ARNGRIM 2, this type of steel was used in both the BBC's Blattnerphone and Marconi–Stille recorders, as well as the Lorenz Stahltone-Bandmaschine.

The steel was produced by melting pig iron, made from pure ores and scrap of known content, in an acid open-hearth furnace. The alloy content was 6% tungsten, 0.7% carbon, 0.3% chromium, 0.2% silicon, and 0.2% manganese. The steel was hot-rolled under controlled temperature conditions into narrow strips, then cold-rolled, annealed roughly 15 times, and finally hardened and tempered. The resulting product was carefully cut into 1000-meter strips, 80 μm thick and either 6 or 3 mm wide.

Uddeholm supplied tape already mounted on reels designed for a half-hour of recording time (roughly 2700 meters of tape). Three lengths were silver-soldered together to make up a reel. Because great care was taken in manufacture, the tape was expensive—around $105 per reel in 1934. In the mid-1930s Uddeholm sold roughly 240,000 meters of tape per year, about 80 thirty-minute reels. In all, Uddeholm sold about 700 reels to various clients between 1931 and 1950, when production ceased.

One other Swedish steel company, Sandviken AB, entered the tape market in 1936. Its tape was similar in manufacture to that made by Uddeholm. The Swedish broadcasting agency Radiotjanst tested this new tape for use with its Marconi–Stille MSR-3 recorders, but decided to continue to use the Uddeholm tape. Sandviken sold a few tapes to the BBC, but its major success came on the German market during the late 1930s, when the company became the major supplier of tapes for the Lorenz Stahltone-Bandmaschine. Sandviken also discontinued production after the second world war, when the Lorenz machines were replaced by more modern equipment.

JAPAN

During the 1930s, Kenzo Nagai carried out a series of experiments with wire recorders at Tohoku Imperial University in Japan. His initial interests were in developing an artificial signal-delay apparatus. To improve the machine's performance, he went on to explore various recording materials, with a primary focus on noise reduction.

Nagai's most important discovery came in 1938, when he, Siro Sasaki, and Junosuke Endo discovered the use of ac bias for sound recording. Some of the benefits of ac bias had been explored, and patented, by Carlson and Carpenter at the U.S. Naval Research Laboratory in 1920. Nagai went beyond Carlson and Carpenter, in that he used ac bias for recording sound, rather than telegraphic signals. Nagai published a description of his work in the journal *Nippon Electrical Communications Engineering* in 1938 and obtained Japanese patent 136,997 on his discovery that same year.

Nagai's innovation is important for two reasons. First, it is the earliest application of ac bias to voice recording, predating Weber and von Braunmühl's work (see Chapter 5) by almost two years and Camras's by over three. There is no evidence that the German researchers saw Nagai's article before their discovery, but the possibility cannot be ruled out. The evidence for influence on Camras is much stronger. The collection of Camras's papers at the Illinois Institute of Technology preserves a list of articles on magnetic recording made in his own hand in 1939. Nagai's article is on that list, and the John Crerar Library in Chicago, which Camras often visited, subscribed to the referenced Japanese journal at the time. When interviewed in 1991, however, Camras denied he had been influenced by Nagai's article.

The second reason Nagai's work is important is that it played a major role in the development of the postwar Japanese consumer electronics industry. Sony selected magnetic recording for development after Sony's founder saw a tape recorder in an American Occupation Forces office. The existence of Nagai's patent allowed Sony and later other Japanese electronics firms to negotiate very favorable terms for access to the patents of the Armour Research Foundation. This advantage aided Japanese firms in entering the magnetic recording market, which they eventually came to dominate.

REFERENCES

Barrett, A. E., and C. J. F. Tweed, "Some Aspects of Magnetic Recording and Its Application to Broadcasting," *Journal of the IEE,* **82,** 265–281 (1938).

Clark, Mark, "The Magnetic Recording Industry, 1878–1960: An International Study in Business and Technological History," Doctoral dissertation, University of Delaware, 1992.

Clark, Mark, "Suppressing Innovation: Bell Laboratories and Magnetic Recording," *Technology and Culture,* **34,** 516–538 (1993).

Lafferty, William, "The Blattnerphone: An Early Attempt to Introduce Magnetic Recording into the Film Industry," *Cinema Journal,* **22,** 18–37 (1983).

Lafferty, William, "The Use of Steel Tape Magnetic Recording Media in Broadcasting," *J. SMPTE,* 676–882, June 1985.

Morton, David, "The History of Magnetic Recording in the United States, 1888–1978," Doctoral dissertation, Georgia Institute of Technology, 1995.

Nagai, Kenzo, Siro Sasaki, and Junosuke Endo, "Experimental Consideration upon the A. C. Erasing on the Magnetic Recording and Proposition of the New Recording Method," *Nippon Electrical Communications Engineering,* No. 13, 445–447, March 1938.

Theile, Heinz, "On the Origin of High-Frequency Biasing for Magnetic Audio Recording," *J. SMPTE,* 752–754, July 1983.

Theile, Heinz, "Magnetic Sound Recording in Europe Up to 1945," *Journal of the Audio Engineering Society,* **36,** 396–408 (1988).

Figure 5–2   The first AEG magnetic tape lab recorder and members of the
tape recorder team, photographed on November 10, 1933: K. S.
Müller (second from the left), Theo Volk, Hans Westpfahl, Eduard Schüller, and Fritz Voigt (the persons on the extreme left and
the extreme right could not be identified. [*Courtesy of AEG /Hans
Westpfahl.*]

was 80 to 100 µm thick, had to be accelerated, kept at constant speed, then
stopped undamaged regardless of its poor resistance to tearing. The team determined that low wow and flutter, to use today's terms, could not be achieved with
a large-diameter, slowly revolving capstan.

Eduard Schüller's Invention of the Ring Head

On August 1, 1933, a 29-year-old engineer named Eduard Schüller joined
the AEG team. His dissertation, "On Magnetic Sound Recording on Steel
Tapes," had qualified him for the job. As an outcome of his postgraduate studies,
Schüller took out German patent DRP 660,377. This patent describes the "ring
head," the basis for all future magnetic recording heads, and one of the most fundamental inventions in magnetic recording (Fig. 5–3). With its rounded and
highly polished surface, the ring head meets the demands of the delicate tape,

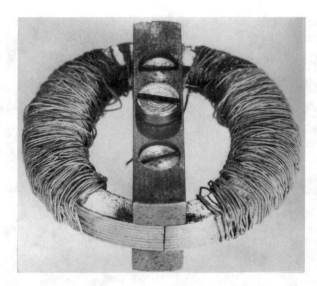

Figure 5–3   The "ring head," designed by
Eduard Schüller (about 1934). [*Courtesy of
AEG/Hans Westpfah*l.]

offering great advantages in magnetic properties over the older, sharpened-head,
design. The magnetic field produced could now be focused precisely onto a very
small region of the magnetic tape. Ring heads soon were built to perform all
three magnetic functions: recording, playback, and erasing. Schüller equipped
the first AEG lab recorder with ring heads, a measure that improved its quality
considerably and, in late 1935, he was appointed the chief engineer of AEG's
magnetic recording work.

Schüller took out his patent on the ring head on Sunday, December 24,
1933, and this was truly a Christmas present for the future of magnetic recording.

The Development of Coated Tape

The weakest feature of the early tapes was their tendency to tear, a defect
detected a few weeks after manufacturing. Eventually Matthias found that embedded carbonyl iron weakened the cellulose acetate considerably. So he decided to manufacture a tape with two layers: a 30 μm thick cellulose acetate base
film, plus a 20 μm layer of carbonyl iron powder mixed with more cellulose acetate to bond chemically with the base film—thus creating a genuine coated tape
according to Pfleumer's model. Although as many as 16 synthetic materials were
tested for suitability as a base film, in the end a recipe similar to that used by Agfa
for safety film gave the most reliable results. In all probability, helpful advice
came from John Eggert, a highly reputed scientist employed at Agfa. Matthias
built a lab machine "similar to those used for the production of photographic
film" and, at the beginning of 1934, tapes were produced by casting 5 mm wide
stripes on a 12 m long endless steel tape. The magnetic layer was about 4 mm
wide, leaving the edges transparent.

Unfortunately, the only laboratory model of the recorder had to stay in Berlin. Therefore Matthias, at the BASF lab in Ludwigshafen, had to find his own way to test the tapes. The job of choosing the most suitable variant of carbonyl iron was given to two BASF physicists, Erwin Lehrer and Friedrich Bergmann. They had to build their own endless-loop test equipment and even magnetic heads based on Schüller's design. The teams in Berlin and Ludwigshafen had to explore basic technologies and experimental techniques that we take for granted today in the design and manufacture of magnetic recorders and tape. Lack of efficient measuring equipment presented one of the first obstacles to be solved by Lehrer and Bergmann. They and other scientists developed measuring devices to establish the important electronic, magnetic, and mechanical quality parameters, which they often first had to define. Their work eventually offered alternatives to purely subjective listening tests.

Improved Laboratory Recorders

Eventually, a second lab recorder was constructed, equipped with a fast-rotating, small-diameter capstan and, of course, Schüller's ring heads (Fig. 5–4). Erasing was achieved by means of a permanent magnet. The ring heads for recording and playback were mounted on a revolving carrier plate so that, during a given operation, only one of the heads came in contact with the tape. The tape speed was 1 m/s, and the quality of reproduction matched contemporary phonograph

Figure 5–4   AEG's second lab recorder, about 1933–1934. [*Courtesy of AEG /Hans Westpfahl.*]

Figure 5–5   AEG's third lab recorder, intended for the launch at the 1934
Berlin radio exhibition. [*Courtesy of AEG/Hans Westpfahl*].

recording for the first time. As a result, AEG decided to launch the magnetic tape
recorder at the 1934 Berlin radio exhibition, entering into competition with steel
wire and tape machines. To this end, a third type of apparatus was designed having
a new tape drive system that resulted in a considerable improvement in sound
quality (Fig. 5–5). About 10 units were to be built for demonstration purposes during the Berlin event.

The Name Magnetophon Is Chosen

At BASF, Matthias now manufactured tapes in lengths of 600 meters,
adding a small amount of nitrous cellulose to the base film, which considerably
improved the breaking strength. The tape was widened to 6.5 mm, but still the
magnetic layer was only 5 mm wide. Protruding coating at the edges had been
found to cause wear on the magnetic heads. Overall, the plastic tapes were superior to Pfleumer's paper tapes, the noise in particular being less annoying. So
Matthias and his team evidently won the competition. On June 1, 1934, a 5000-
meter batch of magnetic tape produced according to AEG specifications was sent
to Berlin. Shortly after that, the chemists produced 50,000 meters for the introduction of the machine at the 1934 exhibition mentioned above. All these tapes,
of course, were manufactured more or less under laboratory conditions. Now the
baby had to be given a name, and AEG and BASF decided to call their "magnetic
phonograph" machine and tape the Magnetophon.

Unfortunately, shortly before the show in Berlin, AEG encountered technical problems—bad hum from the interaction of the recorder's motor, brake

magnets, and electronics—and canceled the machine's debut. They also found that the tape still broke quite easily. It was "back to the drawing board" for both AEG engineers and BASF chemists.

The Fourth Lab Model

AEG built a new laboratory recorder, the fourth and last one (Fig. 5–6). The most unusual feature was the horizontally arranged "turntable" layout in which sheet metal disks rotated with the tape wound on flangeless hubs, making operation more convenient. Playing time was 10 to 15 minutes, depending on the tape thickness, reduced now to about 80 μm. The apparatus consisted of two containers made of black, polished wood, one housing the loudspeaker and the amplifier, the other the tape transport. Dc bias current was preset in the factory. Accompanied by Schüller, one of these machines arrived at Ludwigshafen in mid-April 1935, and he insisted on explaining the operation to his colleagues. This apparatus, it should be emphasized, was the first "genuine" tape recorder the Ludwigshafen team had set eyes on. Test recordings of music and speech demonstrated the progress that had been made.

Figure 5–6 AEG's fourth and last lab recorder used for trial recordings in Ludwigshafen, April 1935. [*Courtesy of AEG Archives.*]

The First Production Tape Recorder: Magnetophon K1

Although the recorder worked well, it overheated after a short time, and AEG put Theo Volk in charge of building a completely new machine. The result was christened the Magnetophon K1 ("K" standing for *Koffer* or portable case). This machine set the pattern for the modern tape recorder with respect to the placement of its reels, heads, and tape path, as well as its three-motor design, the best for tape handling. This transport design, together with further progress in the base film formulation, allowed the tape thickness to be reduced to 50 μm. The tape, 6.5 mm wide, ran past the heads at a speed of 1 m/s. The joint development team decided they would show the improved Magnetophon at the radio exhibition in Berlin in 1935.

By the spring of 1935, BASF chemists put larger scale coating machines into operation, thus moving the Magnetophon tape, called Type C, from development to production. Tape production now was accomplished by casting a 20 mm wide web, coated full width. Afterward, disk-shaped knives cut the web into stripes of the required width, standardized in March 1935 at 6.5 mm ± 0.1 mm. This method led to considerably higher production figures and also solved the problem of the protruding edges. For the Berlin exhibition scheduled for mid-August 1935, the chemists produced 300 km of Type C tape.

THE DEBUT OF THE MAGNETOPHON TAPE RECORDER

The AEG model Magnetophon K1 and Magnetophon Type C tape were a sensation at the Berlin exhibition that summer of 1935 (Fig. 5–7). People were amazed to be able to hear their voices an instant after recording. The demonstration was overshadowed, however, by a fire midway through the show that destroyed the hall in which the prototype machine and tape were housed. The AEG team across town worked day and night and produced a second prototype K1 tape machine that was shown during the final days of the exhibition in a makeshift booth. In a letter to Dr. Gaus in Ludwigshafen, Matthias described the event:

> The Magnetophon machine created a great deal of interest at the exhibition. The dealers recognized the fact that this machine was the "hit" of the show, something we heard as well. They realized the many important applications of the machine, including its use as a dictation and news-gathering device, as well as the perfect replacement for phonograph records. Many inquiries about the delivery of 25 or 50

Figure 5–7   Magnetophon K1 and Magnetophonband Type C, as launched at the Berlin radio exhibition of August 1935. [*Courtesy of AEG Archives.*]

machines at a time were heard. Many people indicated their desire to see variations of the Magnetophon: combined with a telephone; an auxiliary unit for a radio; prerecorded music and entertainment tapes; a special version to create artificial reverberation for open-air concerts. (Matthais, 1935)

The enthusiastic report about the Magnetophon at the Berlin show, written by the 39-year-old Friedrich Matthias, combined with the dedication of the 31-year-old Eduard Schüller and his AEG team to replace the burned-up prototype before the show ended, demonstrated the high value that they and others placed on the invention. After only two and a half years of cooperative development work by AEG and BASF, the promising system was ready. The price was attractive at 1350 Reichsmarks. For comparison, the most advanced steel wire dictating machine, the Textophon from Lorenz, cost RM 1365. An hour of tape cost RM 36, about one-third the price of steel wire for the Textophon, and one-seventh that of steel tape (RM 260/h) for the competing professional Lorenz Stahlton-Bandmaschine recorder. The monthly salary of a trained technician in those days was about 250 Reichsmarks. Clearly, the achievable recording technology was beyond the means of most consumers; rather, it was aimed at the professional audio and government markets.

THE MAGNETOPHON GOES INTO PRODUCTION

Nevertheless, the sales outlook of the new recorder and recording medium in various markets looked bright. BASF geared up to make tape in volume by ordering a full-scale production machine from the Dresden firm Köbig, one of the long-time suppliers of Agfa. Delivered in late 1935, the modified photo paper manufacturing machine was subjected to an extended series of test runs. Regular production started in June 1936.

The basic production technique remained unchanged, but was on an enlarged scale. The production machine consisted of a narrow cabinet containing a continuous copper belt, 28 meters long and 60 cm wide. Liquid cellulose acetate was spread onto the belt and allowed to dry partially. In the final step of the tape-making process, the treated sheet of cellulose acetate was coated with a viscous mixture of carbonyl iron powder, suspended in a cellulose acetate–acetone solution. Then the web was run about 12 meters through a hot air channel to dry it to some extent. After passing 26 meters through the hot-air channel, the finished tape was dry enough to leave the support of the copper belt. It was then wound into a wide spool of finished bulk tape, ready for slitting into 84 tapes, each 6.5 mm wide. The tape base film was 30 μm thick, with a 20 μm coating, close to the specifications of modern professional audio tape.

In the mid-1930s, the speed of the tape manufacturing process was limited to only 3 m/min. A thousand-meter run required 6 hours. (A modern tape plant would produce a 1000 meters in little more than 2 minutes.) Nevertheless, this system represented the first true magnetic tape production machine worldwide (Fig. 5–8).

Magnetite Replaces Carbonyl Iron

Between the spring and summer of 1936, shortly after putting the coating machine into operation, the BASF chemists replaced carbonyl iron with magnetite. This type of iron oxide, Fe_3O_4, had been mentioned by Pfleumer in his 1933 patent. It offered a higher coercivity and greater remanence, the two main attributes of a recording medium. The fine particles, with lengths of about 1 μm, produced lower noise, and the abrasivity was diminished. Instead of light gray, the Magnetophon tapes were now black (Fig. 5–9).

The first formal test of the new magnetite-based tape came in November 1936, when BASF engineers (probably Matthias and his electrical engineer, Paul Friedmann) recorded a concert by Sir Thomas Beecham and the London Philharmonic Orchestra in the "Feierabendhaus," the employees' auditorium

Figure 5–8 The magnetic tape production machine, put in service in 1936. The caster for the magnetic layer is on the right (marked by the vertical framework). The machine was approximately 16 meters long and had a web width of 65 cm. [*Courtesy of BASF Archives.*]

Figure 5–9 Magnetophonband "Type C" (cellulose acetate coated with Fe_3O_4), and its cardboard box, appearance from 1936: tape 1000 meters long is wound on a 70 mm sheet metal hub. [*Courtesy of BASF Archives.*]

on the factory grounds. Understandably, the poor-quality dc bias electronics failed to produce a recording as good as those available from the competing wax transcription disks used by recording companies and radio stations in Germany.

From Magnetophon K1 to Magnetophon K4

In the fall of 1935, the magnetic recording department of AEG moved to a more spacious building, which became its home for the next 10 years. After building 10 to 15 Magnetophon K1 models, which ran tape at 1 m/s, the company introduced the K2 models in 1936, most of them with the slower speed of 0.77 m/s. The playing time thus increased to 22 minutes with a 1000-meter tape length. Planned or by chance, the new speed was almost exactly half the 1.5 m/s speed of the still-competitive steel tape magnetic recorders. Around 100 K2s, and the successor K3s, were built in the next two years. Then, in 1938, AEG introduced its first commercially successful tape recorder, the Magnetophon K4, with its modern amplifiers and modular magnetic head stack, the first Magnetophon to offer play-while-record monitoring. The K4 was the best of the dc bias machines.

The Magnetophon Pays a Brief Visit to America

In November 1937, the AEG affiliate in Schenectady, New York, received a Magnetophon (probably the K3 version) to show to colleagues at the General Electric Company. The German company had wanted to convince the GE marketers to sell the tape recorder in the United States, but the only result was a precisely written technical report completely ignoring the potential of the new technology. To summarize, there were three main objections: poor dynamics, poor frequency response, and too many parts. It was also stated that "radical changes in design would be necessary"—in other words, "not invented here."

After the end of the second world war, it was common opinion in the United States that all information about magnetic tape recording was treated as a German state secret—despite the Magnetophon sent to GE, the many detailed publications in the German trade press, and newspaper accounts published throughout the war.

A Promising New Oxide: Gamma Ferric Oxide

By 1939 the state of the art of tape quality had progressed far beyond its 1935 debut. Probably the most important progress came that summer with the introduction of gamma ferric oxide tape, γ-Fe_2O_3, a reddish-brown iron oxide. This material, dating back to a 1935 BASF patent, was to prove so effective that it was not until 30 years later, when chromium dioxide was introduced in 1971, that anything fundamentally better would replace it. Tape formulas in Germany, Japan, and the United States in the 1940s, 1950s, and 1960s, were evolutionary improvements on this BASF patent.

The γ-Fe_2O_3 magnetic tapes did not exhibit the erasure problems of magnetite (namely, an "erased" recording that reappeared at a clearly heard level days later). Also, the γ-Fe_2O_3 particles were considerably smaller than the Fe_3O_4 particles, and on dc biased Magnetophons, the new tape achieved a signal-to-noise ratio of a little more than 40 dB, and a frequency response from 50 Hz to 5 kHz— quite impressive, but still unacceptable for true broadcast quality.

Between 1935 and 1943, the Ludwigshafen chemists used three magnetic formulations for the tape sold by the brand name "Magnetophonband Type C":

- Up to the summer of 1936: carbonyl iron (light gray, metallic pure iron)
- From the middle of 1936 to the summer of 1939: Fe_3O_4 (black, cubic iron oxide)
- From the autumn of 1939 to 1943: γ-Fe_2O_3 (gamma-type red, cubic oxide[*])

Throughout the development of these magnetic materials, the base film remained the same—cellulose acetate, hence the designation "Type C." Although polyvinyl chloride (PVC) and later, polyester (Mylar) base films eventually replaced cellulose acetate, the technology lasted well into the 1960s. Type C tape had advantages in manufacturing. The requisite acetone was cheap and plentiful, and the cellulose acetate was used in both parts of the tape, base film and magnetic coating, resulting in a cost-effective and straightforward manufacturing process. Acetate tape had its weaknesses—it broke easily and was sensitive to humidity. However, it had one advantage over its plastic successors in that it broke cleanly and did not stretch. Thus the recording engineer could easily splice broken tape back together with no audible break in the sound.

THE MAIN USERS OF THE MAGNETOPHON

By the end of 1937, around 200 Magnetophons were actually in use, most of them apparently models FT 1 to FT 4 office dictation systems, based on the same transport and electronics as the Magnetophons. Other users in Germany included

---

*Needle-shaped variations were developed at the beginning of the 1950s using the same formula.

broadcast, military, some government authorities, and the postal service. Later on, government authorities would use K4 machines as eavesdropping recorders to monitor domestic and diplomatic telephone calls. In 1945 a pair of these K4s would become star performers when John T. Mullin chose them to open the age of magnetic tape recording in the United States.

The German military in the 1930s naturally adopted any technology that would advance its cause, and there were many uses for magnetic recording. Portable recorders were developed and used in a variety of field applications. The military versions of the AEG recorders were called Tonschreiber, or "sound writer." The broadcast machine R23 was modified and dubbed Tonschreiber D. Other models appeared at the same time, including the Tonschreiber C (Caesar), a tiny, spring-driven machine, the most portable magnetic recorder yet created. The most interesting and technically advanced prewar model was the Tonschreiber B (Berta), a transportable deck and amplifier with variable recording speeds from 90 mm/s to 1.2 m/s. In addition to the normal head stack, the field recorder featured an additional four-gap spinning head assembly for playback at variable speeds without a change in pitch. In other words, this was a way to compress or expand speech, or to encode and decode telegraphic messages.

In the mid-1930s, the German radio service, the Reichs-Rundfunk-gesellschaft (RRG), used several different sound recording systems. In terms of quality, the wax disk, which could record for only 4 or 5 minutes, was unsurpassed. Since, however, it was difficult to handle, its use was limited to music recordings. Cellulose acetate disks in huge numbers were used as carrier media for on-the-spot reports, and the Lorenz steel tape sound recorder made field recordings. Even a modified version of the optical sound film system was to be found. After having tested various developmental AEG tape recorders between 1935 and 1938, the RRG declared the Magnetophon to be one of its future recording devices. They recognized the cost and performance advantages of magnetic tape—including the ability to record uninterruptedly for 22 minutes—over the wax transcription disk. At first, the RRG used dc bias Magnetophons for noncritical recordings, such as speeches and interviews, because, as was to be expected, the RRG engineers were not entirely pleased with the dc biased Magnetophon's recording performance.

The Magnetophon's Recording Performance

The first implementation in German radio stations was the R22, based on the AEG Magnetophon K4 tape transport, but with special amplifiers developed by RRG engineers (Fig. 5–10). The broadcasters realized the importance of the new Magnetophon for electronic news gathering (ENG) operations and, working with AEG, helped develop the Magnetophon R23 in 1939. This compact, battery-

Figure 5–10 Magnetic tape recorder "R22" of about 1940, as used by the RRG, the German radio broadcasting service. Eventually, most of these machines were converted to ac bias. [*Courtesy of AEG/Hans Westpfahl.*]

powered portable deck still retained the dc bias recording circuit, with its inferior sound quality.

THE BIG LEAP IN QUALITY: AC BIAS RECORDING

The director of the RRG electronics development laboratory, Hans-Joachim von Braunmühl, was sure that the Magnetophon could be improved enough to record music, so he put RRG engineer Walter Weber to work perfecting the new tape technology. The RRG magnetic tape lab became, as it were, the third branch of the Magnetophon R&D effort, but operated independently of the AEG operation across town.

In April 1940 Walter Weber rediscovered one of the most important techniques in magnetic recording history. He discovered ac bias through a combination of systematic research and a bit of luck. He had been experimenting with phase cancellation circuits in an attempt to reduce the distortion and noise of dc bias recordings. An amplifier in a test setup went into oscillation, accidentally creating an ac bias current in the recording circuit. After some systematic engineering detective work, Weber discovered what had happened and learned to recreate the phenomenon. Negotiations with AEG were made to establish patents and licensing for the Magnetophon. The most important of the resulting patents, DRP 743,411, was issued December 24, 1943, exactly 10 years after Eduard Schüller took out his ring head patent.

Weber was evidently unaware that Kenzo Nagai and coworkers in Japan had discovered ac bias two years earlier and had described their findings in a paper published in 1938. Ironically, Lehrer in Ludwigshafen had experimented with ac bias even earlier while testing magnetic materials in 1934, but was unable to make the process work well. At AEG, Schüller also unsuccessfully tested ac bias but did not communicate with Lehrer.

Thus Weber was not the first to apply ac bias to magnetic recording, but he deserves great credit for recognizing immediately the practical value of his discovery and applying it to improve the Magnetophon's recording quality. The Magnetophon now had the best audio quality of any recording technology in the world, far better than the old dc bias tape recording, but also better than the current commercial competitors to Magnetophon tape.

The superiority of the new "high fidelity" magnetic recording process was made clear in the Berlin exhibition on June 10, 1941, in an AEG-RRG demonstration publicly reported by the local press: *"A fantastic experience in electrical sound recording, a total revolution in sound recording."* No wonder the journalists were impressed: for the first time in history, a sound recording had achieved 60 dB dynamic range and a frequency response of 50 Hz to 10 kHz (figures based on weighting filters and tolerances of the day).

AEG-RRG researchers soon added stereo recording to their 1940 high-fidelity Magnetophon triumph. As early as 1942, they made test recordings of the Berlin Philharmonic Orchestra using three Neumann condenser microphones and a stereo version of the R22 studio deck.

The next stage was to improve the tape handling, wow, and flutter of the K4-based Magnetophons. The breakthrough was Model K7, the first tape deck with synchronous motors, introduced in the spring of 1943. Subsequently, for the first time, even stereophonic *"acoustically accurate"* sound reproduction was demonstrated to the public. Since this was before the era of stereophonic radio broadcasting, all these recordings went to the radio archives labeled "for archival purposes only." The stereo version of the Magnetophon never made it into regular production, and wartime pressures and postwar chaos ended serious stereo experimentation until the late 1940s.

The German film industry took an immediate interest in the new high-quality sound recording method. Color film had recently been introduced, and optical sound-on-film suffered from the new color emulsions, which did not reproduce sound as well as black-and-white film. In 1941 engineer Karl Schwartz of Klangfilm GmbH in Berlin received patent number DRP 969,763 for his solution to the fundamental problem of film–projector transports: the use of perforated magnetic tape (oxide coated on clear film stock).

WARTIME DEVELOPMENTS

At the end of July 1943, the BASF Magnetophon tape plant in Ludwigshafen was completely destroyed by fire following a non-war-related, accidental explosion of a tank car. The disaster wiped out all the tape manufacturing machinery in the factory, including the only coating machine, marking the end of "Magnetophon Tape Type C" production in Ludwigshafen. Quick improvisation was necessary to meet RRG tape needs, as well as wartime military tape consumption, and an alternative manufacturing site had to be located. Despite difficult circumstances, the tape makers found three answers to the problem.

As mentioned earlier, BASF engineers had periodically discussed magnetic tape manufacture with scientists working for Agfa in Wolfen. Initially, the Agfa scientists took a skeptical view of the Magnetophon developments. Their approach changed dramatically after the glorious Berlin demonstration of the ac bias Magnetophon in June 1941. Farsightedly, the Agfa movie specialists correctly predicted magnetic tape's potential and, under the direction of John Eggert, started R&D activities on magnetic tape with assistance from Ludwigshafen.

In late 1942, Agfa had already begun manufacturing test runs of the Type C formula developed in Ludwigshafen. By the end of February 1943, they had sent four samples 50 to 70 meters long to BASF Ludwigshafen for analysis. In the wake of the explosion in Ludwigshafen a few months later, Eggert and his team found themselves compelled to take responsibility for all Magnetophon tape deliveries. All the BASF oxide, which Ludwigshafen could still produce in volume, as well as documentation of magnetic tape production know-how, were sent to Wolfen. The operation was the beginning of Agfa's commercial tape production. Two old Köbig photo paper machines, dating back to 1914, were reactivated (amazingly, they were in use up to 1963) and used a modified form of the Ludwigshafen cellulose acetate tape production process. RRG, highly interested in promoting good-quality Wolfen deliveries, assisted in checking the tapes. In 1944 Wolfen produced 31,300 km of Type C tape.

A New Tape Base Material: PVC

The second option available to permit continued tape manufacturing, this time in Ludwigshafen, was a change in base material from cellulose acetate to a new plastic, polyvinyl chloride or PVC, which had been developed and manufactured since the end of the 1930s under I.G. Farben's trade name of Luvitherm. BASF researcher Heinrich Jacqué, inventor of the Luvitherm manufacturing process, started production in September 1940. As early as 1938, Jacqué had suggested using the method to produce trial tapes, replacing cellulose acetate, and putting the oxide directly into the PVC mixture and rolling the result by means of a calendering machine into 50 μm thick homogeneous tape webs, ready for slitting.

When the tape plant burned down, BASF realized that the Luvitherm method eliminated the coating process, avoiding the need to replace the destroyed coating machinery. Jacqué used the calendering machine to make test runs of the new tape, called "Magnetophon Type L" (the "L" standing for Luvitherm). The BASF and AEG people were pleasantly surprised to discover that the new Type L tape performed far better than the old Type C product, with an amazing 10 dB improvement in signal-to-noise ratio. The downside was that the tape's sensitivity also decreased by 10 dB, a special handicap for tube amplifiers. About 50 of these tapes would form the basis for the introduction of magnetic tape in the United States. Increasing air raids, however, forced the removal of the calendering machine to Gendorf, Lower Bavaria, in the autumn of 1944.

In 1942 BASF chemist Rudolf Robl was given the job of finding a way to coat iron oxide onto the new Luvitherm PVC base film. He had to develop a new coating process that differed from the manufacture of the original Type C tape. Acetone, the solvent used to make Type C tape, dissolves PVC readily, so Robl had to find a more suitable solvent. Coincidentally, BASF had begun producing tetrahydrofuran (THF), which Robl discovered could serve as a solvent in the PVC coating process.

Unlike the manufacture of Type C tape, in which base film production and coating occurred in one long process, Robl had to coat premanufactured webs of PVC base film. The long webs of PVC were difficult to handle during the coating process. Robl's solutions to materials handling formed the basis for tape manufacturing methods still in use today. Robl also suggested the mixing of titanium dioxide into the base film to help the end user more easily distinguish the back and front of the tape. He called his new tape Type "LG" ("G" for the German word *Guss*, for coating). Type LG was thoroughly modern—a plastic base film with a γ-Fe_2O_3 oxide coating. By the end of 1943, Robl was ready to go into production with the new Type LG tape.

Because of air raids, BASF set up Robl in late 1944 in a former truck garage in Aschbach, about 50 km east of Ludwigshafen. Starting in early 1945, Robl and his five employees produced 1600 km of Type LG tape per month. Ludwigshafen could still deliver iron oxide (and did so in 50-liter milk churns). The base film came from the plant in Gendorf. The Aschbach plant stayed in operation until 1947, two years after the end of the war.

AEG and BASF Launch Magnetophon GmbH

AEG and BASF cofounded "Magnetophon GmbH," or Magnetophon Company, in 1942 in Berlin. One of the two corporate officers was Friedrich Matthias. In the summer of 1944, he and his small team moved to Waldmichelbach, 2 km from Aschbach. Waldmichelbach would be the tape-slitting and packaging center for all tape manufacturing. Those locations shipped their wide rolls of manufactured tape to Matthias's new facility, where they were slit into 6.5 mm wide tape, wound onto hubs, packaged, and sent to end users. Since both Gendorf and Wolfen are located around 360 km from Waldmichelbach, the wartime transportation and logistical problems Matthias faced are readily imagined.

POSTWAR DEVELOPMENTS

Wartime conditions—air raids, loss of qualified staff to the army—forced curtailment of production at the AEG Magnetophon plant for all but military needs. Late in 1944, the main facility was bombed heavily, and production came to a complete stop.

After the end of the war, Eduard Schüller and his crew were moved to the region north of Hamburg and, in June 1946, the Hamburg office of Magnetophon GmbH was operative again, mainly repairing and checking recorders used by broadcast stations. About 1948, production of Magnetophon K8 recorders started in the small town of Wedel, north of Hamburg, under the direction of Schüller. He moved to Berlin in 1950, where he became head of the basic research department of Telefunken (a subsidiary of AEG since 1941). Telefunken launched the first German home magnetic tape recorder in 1952. In the same year, studio recorder Magnetophon T9 was introduced, a completely new design. It became the most popular "workhorse" in German broadcast stations and sound studios during the next decade.

After the end of the second world war, Germany was divided into four zones of occupation. As chance would have it, the magnetic tape production plant in Ludwigshafen was in the French zone, Waldmichelbach and Gendorf facilities were in the American zone, and Wolfen was in the Russian zone. Only the British zone depended on deliveries from the other zones, a situation to be altered in 1948.

A year after the end of the war, in 1946, the BASF crew began to rebuild the tape-making facility in Ludwigshafen. About 30 employees battled hunger and difficult circumstances to set up a tape plant in a makeshift building (Fig. 5–11). The main customers were German broadcast stations. The team's biggest problem was to find enough PVC base film. The calendering machine in Gendorf (American zone of occupation) could not be transferred back to Ludwigshafen (French zone). Immediate postwar Allied politics impeded intra-German trade across zones of occupation. Even the machinery in nearby Aschbach, also in the American zone, was not available to the BASF team, so they had to overcome postwar shortages by scrounging parts to rebuild the tape plant in Ludwigshafen. Necessity is the mother of invention: digging through the ruins of the old tape plant that

Figure 5–11   Visual testing of Magnetophon tapes in Ludwigshafen, around 1948. [*Courtesy of BASF Archives.*]

had been destroyed in 1943 yielded two old slitting machines that were refurbished (Fig. 5–12).

As already mentioned, in 1944 the calendering machine used to produce both the BASF homogeneous tape and PVC base film was moved to Gendorf in lower Bavaria. After the war, around 1948, Friedrich Matthias set up a complete tape manufacturing plant within the "Anorgana Gendorf" in the American zone, producing tape with the brand name Genoton. As a result, the Ludwigshafen people felt obliged to break off all economic contacts with Gendorf, and that meant also doing without the precision PVC base film. The Genoton plant operated until 1956, when it was closed and some of its personnel were transferred back to BASF in Ludwigshafen. Matthias died in 1956 near Gendorf.

After the war, the Wolfen plant, located in the future East Germany, became a state film and tape factory eventually called "Orwo." Production figures rose from 8000 km of tape in 1945 to 300,000 km in 1953, when two new coating machines of the Köbig type were installed, replacing the 1914 models. Beginning in 1946, the neighboring Agfa Farbenfabrik delivered the iron oxide, a job taken over by the tape makers in 1952. Up to 1954, improved versions of the Magnetophon Type C under the brand name Orwo were manufactured, followed in 1954 by the Type CH for low-speed home recorders. The Wolfen tape plant was relocated to Dessau in the following years. Production continued until the early 1990s.

Figure 5–12 Slitting tape on a machine dug out from the debris of the first magnetic tape plant, around 1948. [*Courtesy of BASF Archives.*]

As mentioned above, three of Germany's four zones of occupation had magnetic tape production. Nevertheless, the broadcast stations' great demands already exceeded the production capacity. This in particular applied to the Nordwestdeutscher Rundfunk (NWDR) in Hamburg and Cologne. In all probability, the British authorities encouraged the West German Agfa branch to build a new tape factory in Leverkusen, near Cologne, on the premises of the Bayer works (the home of Bayer aspirin). Be that as it may, in October 1947, the first Agfa tape specimens were successfully tested at the Hamburg Radio labs. Almost traditionally, a Köbig machine was adapted, and subsequently NWDR ordered some 2500 tapes, based on cellulose acetate and coated with $\gamma\text{-}Fe_2O_3$.

The Agfa production conditions apparently were better than those in Ludwigshafen, giving the Leverkusen tapes an edge in competition. The Bayer works at nearby Uerdingen produced suitable iron oxide. Scientists from Wolfen in 1948 joined the team, and the result in 1949 was the well-known Agfa Magnetonband F. In 1950 the daily capacity amounted to 150 thousand-meter tapes. At the same time, magnetic film "MF" went into production and was delivered to the upcoming German film industry, soon to be followed by the improved versions MF 2 and MF 3. In 1950 the production range was extended to Agfa Magnetonband FS for the "slower" speeds [i.e., from 0.38 m/s to 9.5 cm/s (15 to 3.75 in./s)].

THE BIRTH OF THE CONSUMER MARKET

The beginning of the 1950s saw the appearance of the first consumer tape recorders on the German market. Private consumption of tape was on the up-swing, although the largest consumers of German-made magnetic tape were still the German and Allied Forces radio stations. Consumers needed tape that would play with acceptable quality at speeds slower than 19 cm/s (7.5 in./s). To meet the need, BASF developed a tape with increased sensitivity, Type LGH. With increasing sales of home recording decks, the consumer tape market rapidly eclipsed the professional tape segment in size.

BASF accepted the fact that an American company, 3M, had made certain technical advances in tape manufacturing by naming its 1953 consumer tape series Magnetophon Tape Type "LGS." Only industry insiders knew that the "S" designation meant "Scotch-compatible," indicating the technical compatibility of Type LGS with the new 3M formulation being used on many consumer tape decks of the time.

From its rebirth in America, magnetic recording spread rapidly around the world in radio broadcasting, in consumer applications, in professional and military data recording, including computers, and in the motion picture industry. Since the postwar Allied Commission had invalidated most German patents and therefore the rights to royalties on their prewar and wartime inventions, AEG, BASF, and Agfa received no immediate benefit from the tremendous world-wide growth of their inventions. Patents filed after the war by the German companies, and by European newcomers such as Studer/Revox and Kudelski in Switzerland, and EMI in Britain, later reestablished Europe as a place of magnetic recording innovation. Japanese companies also began to play an increasingly important role in postwar magnetics, with Sony Corporation's entry into the field in the late 1940s, followed by Matsushita (Panasonic), Toshiba, and others.

SOURCES

This chapter is based on the internal and external correspondence of the BASF departments concerned with the Magnetophon tape (including, in particular, the correspondence with AEG in Berlin) and a number of recently accessible, important documents on the information exchange concerning magnetic tape between BASF Ludwigshafen and Agfa Wolfen, and the beginning of tape production there (by courtesy of the Wolfen Film Factory Central Archives). The following provided information on the history of Magnetophon tape and recorder production from about 1932 until the mid 1950's:

- Two "keepsake albums" arranged by Hans Westpfahl, a member of the Magnetophon development team (Figs. 5–2, 5–4, 5–5, and 5–10, in particular).
- Personal documents related to Dr. Walter Weber's work with the Reichs-Rundfunkgesellschaft, by courtesy of Dr. Jörg Weber, Berlin.
- Speeches by Dr. Rudolf Robl and Heinrich Jacqué on November 29, 1959 (internal BASF celebrations to mark the 25th anniversary of the Magnetophon tape).
- Dr. Karl Holdermann and Dr. Rudolf Robl, Magnetophon tape operation, unpublished manuscript, 1956.
- Dr. Paul Zimmermann, "Magnetic Tapes—Magnetic Powders—Electrodes," Ludwigshafen, 1969; in particular the contribution by Dr. Friedrich Bergmann: "Magnetic Powders." These sources give the history of carbonyl iron at BASF and describe some aspects of the early days of magnetic tape production.
- Friedrich Matthias report dated 9/4/35 to Dr. Wilhelm Gaus. BASF Cenral Archive.

6 Building on the Magnetophon

Beverley R. Gooch

Before the second world war, the magnetic recording industry in the United States was almost nonexistent. During the war years, there was an upsurge in magnetic recording driven by the demand for recorders for military applications. As described in Chapter 4, all the recorders produced in those years used wire or steel tape media and were developed mainly by two companies: the Brush Development Company, led by Semi Begun, and the Armour Research Foundation, led by Marvin Camras.

In contrast to what was happening in the United States, prewar tape recorder development in Germany was advancing at a rapid pace. Before the outbreak of the second world war, a German firm, AEG, demonstrated an early Magnetophon tape recorder to General Electric in hopes of securing a license agreement. GE failed to recognize the potential of the early Magnetophon and put more faith in wire recording at that time. Another reason for the failure of the large U.S. electronics firms to show an interest in the German developments can be found in the way American broadcasting was structured in the late 1930s and early 1940s. The radio industry evolved around a highly developed infrastructure of telephone lines that distributed programming to local affiliates. Any recording of this program material was looked on as piracy of the network's property, and the affiliate stations were prohibited from recording network programs. This view of the ownership of content severely limited the need for local radio stations to own magnetic recorders. Phonograph records were used for local programming material and, if recording was required, the refined technology of phonograph transcription was used.

After the United States entered the war, most avenues of exchanging technical information with Germany were closed. The great strides that were made in magnetic tape recording during the war years became known only when Allied military personnel obtained access to the Magnetophon recorders and magnetic tapes and disseminated this information to industry through reports of the Allied intelligence services. The U.S. Alien Property Custodian held all patents on the Magnetophon; any American company could obtain a license to produce the German equipment free of charge upon application. Large U.S. corporations still showed little interest, and the exploitation of the Magnetophon technology rested with start-up companies like Ampex and Rangertone.

After the war, a new magnetic recording industry based on particulate tape media began to develop in the United States and overseas. This was largely due to the transfer of technology from Germany and the war-driven research in the United States. Brush was the first company to offer recorders using magnetic tape to the consumer market. It was soon followed by Ampex, Rangertone, and Magnecord, which developed tape recorders for professional applications. This chapter highlights the development of both consumer and professional magnetic tape recorders from the period immediately after the war to the mid-1950s.

COATED TAPE DEVELOPMENT IN THE UNITED STATES

Early Experiments

The first magnetic tape development program in the United States was initiated by Brush in August 1943 as a part of a contract with the federal Office of Scientific Research and Development (OSRD). Semi Begun of Brush had found that the output from ring-type recording heads was more dependent on the magnetic fields on the surface of the media than on the interior. Thus it was concluded that a recording medium with a thin layer next to the head would yield a better performance than a thick solid medium such as a wire. Based on this experimental evidence and intuition, Begun was convinced that wire and metal tapes would eventually be replaced by recording media having a thin magnetic layer that was flexible and lightweight. Begun was also aware of the work of Fritz Pfleumer, the Austrian inventor who made his first magnetic tape in 1929 and collaborated with the German electronics giant AEG, and later the BASF group of I. G. Farben, to develop paper-based magnetic tape for the Magnetophon. Most of this work was done after Begun left Germany, and he had little specific knowledge of the BASF achievements in tape development. Access to the later German work or technical papers was limited, which meant that most of the tape-making technology had to be reinvented. To help with this task, Begun solicited Battelle Memorial Institute, known for strength in materials science and chemistry, to participate in a joint effort to develop a magnetic particle coating for a recording medium.

The early tape experiments at Battelle were performed by Gerard Foley. His first attempts at making tape particles were very crude; he collected filings from a material known as Alnico, an aluminum–nickel–cobalt alloy, which was used to make permanent magnets for loudspeakers. These filings were milled to reduce their size and suspended in a lacquer solution. The particle solution was coated on a 0.003-inch-thick cellulose acetate base film, which was chosen because it was easy to coat and bonded well with the lacquer. Since the coating was nonuniform and the particle size had quite a spread, the signal output exhibited large variations (10 dB), poor high-frequency response, and high noise.

As the work progressed, Foley tried natural powdered magnetite. While there was some improvement, the variation in particle size was still too great to yield an even coating and produce acceptable performance. Foley then discovered that paint pigments were made from artificially produced magnetite particles, which were also magnetic. Experiments with these particles revealed magnetic and physical properties that were much more consistent than those from the natural magnetites. By mid-1945, using tapes made from the artificial magnetite particles, the total output variations had been reduced from 10 dB to 6 dB and a dynamic range of 35 dB had been achieved. This performance was better than the average wire media and was considered acceptable for Brush's postwar products.

Brush did not have experience or facilities to coat tape in the production quantities needed to support its efforts. The company attempted to interest Eastman Kodak, Meade Paper, Minnesota Mining and Manufacturing (3M), and Shellmar (a maker of printed wrappers for bread), which were experienced in large-scale coating techniques. Shellmar, the only company interested in working with Brush to produce tape, entered into an agreement to supply a paper-based magnetite particle tape in production quantities in September 1945. Experimental production was under way by the end of the year, and by spring 1946, Shellmar was supplying relatively large quantities of magnetic particle media for the Mail-A-Voice disk machine and the BK-401 Soundmirror tape recorder. The 3M contact was significant in the development of magnetic tape in the United States because, after the initial discussions with Brush, 3M became interested in magnetic tape and went on to establish a magnetic tape development laboratory.

Tape Improvements

Following the early tape development at Brush–Battelle, 3M, Audio Devices, Indiana Steel Products, and Marvin Camras of Armour Research became increasingly active in tape development. Moreover, the U.S. access to technical information relating to tape development that had occurred in Germany during the war further promoted the rapidly developing U.S. tape industry. Out of these concentrated efforts, better tape particles began to emerge. The ability to produce and control particles with acicular, or needlelike, shape was a significant step toward better quality tapes. The early Brush magnetite particles were basically cubic in structure and were characterized by low coercivity [100–175 oersteds (Oe)] and retentivity. Low coercivity and retentivity typically limit the high-frequency response and dynamic range of the media. The size and shape of acicular oxide particles made it possible to achieve greater particle density and a higher degree of magnetic orientation than could be obtained with the early cubic magnetite particles. This, in turn, resulted in higher coercivities (250–350 Oe) and retentivity, thus providing substantial gains in dynamic range (60 dB was attainable) and high-frequency response. Subsequently, the use of these particles made it possible to reduce the tape speed while maintaining the same performance, resulting in longer playing time per reel.

In 1947 3M introduced its first commercial tape (Type 100), which used black oxide acicular particles applied to a paper base. The coercivity was approximately 350 Oe, and the product exhibited increased output and improved high-frequency response over the earlier Brush tapes. However, the black oxide tapes were difficult to erase because of a memory effect in which residues of earlier recordings could be heard in the background of subsequent recordings. With the introduction of red acicular gamma ferric oxide (γ-Fe_2O_3), this memory effect was eliminated. In 1948 3M introduced the first acicular gamma ferric oxide tape (Type 101), coated on a paper base. The coercivity was between 250 and 275 Oe. Type 111, which soon followed, used the same particles but was coated on a new cellulose acetate base. The ubiquitous 111 tape became one of the most widely used tapes for professional applications in the 1950s. By 1956, gamma ferric oxide particles were used on all audio magnetic tape manufactured in this country and were gaining use abroad.

With the developments at 3M and by others, Brush was losing its competitive advantage. To stay in the market while improving product performance, Brush continued its alliance with Shellmar to produce the paper-based magnetite tape. However, Brush's strengths were not in chemistry and coating techniques, and the firm was overwhelmed by the much larger research capabilities of the expanding tape industry. Brush stopped trying to compete and began buying tape from 3M packaged with the Brush label and, by 1950, was out of the tape business altogether.

The expanding use of magnetic recording greatly increased the demand for magnetic recording tape, and more companies were eager to capitalize on the growing markets. By 1956 there were six major U.S. manufacturers of magnetic tape: 3M, Audio Devices, Reeves Soundcraft, Technical Tape Corporation, Ferro Print, and Orr Radio Industries. In 1959 Orr Industries merged with Ampex to form the Ampex Magnetic Tape Division. The magnetic tape field became very competitive, a development that gave rise to rapid technological advances and improvements.

CONSUMER RECORDERS

In late 1944 Brush and Armour began to develop strategies to take advantage of the postwar markets. Armour devoted its energies to becoming a major magnetic recording research center and technology licensing operation. Brush's short-term plan was to capitalize on the wire recording technology it had developed during the war. The long-term plan was to manufacture recorders that would exploit the particle media developments made by Brush–Battelle. The implementation of these plans began in 1945 when Brush brought three wire recorders to the market. However, the wire recorders played only a minor role in Brush's postwar business. The company's major thrusts were the particulate tape recorders, the Mail-A-Voice and the Soundmirror.

The Brush Mail-A-Voice

The first recorder to use particle tape media, the Brush BK-501 Mail-A-Voice, was introduced in October 1946. This novel product resembled a phonograph record player (Fig. 6–1). A paper disk coated with particle magnetic material was mounted on a small turntable that rotated at 20 rpm, giving a playtime of 3 minutes. The record/play head was mounted on a standard phonograph arm,

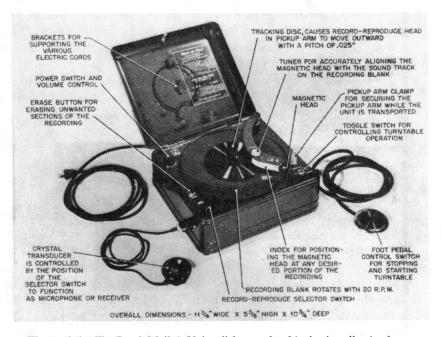

Figure 6–1   The Brush Mail-A-Voice disk recorder. [*Author's collection.*]

which was guided across the magnetic disk by a spiral guide mechanism in the center of the turntable. The disk was 9 inches in diameter, and the spiral sound track started with an inner diameter of 5 inches and ran to the outer edge of the disk. The spiral track width was 0.014 inch, with a pitch of 0.025 inch. Ac bias was used for recording, but the medium was erased by a permanent magnet. The user could record short messages on the disk, then fold and mail it. This product had an enthusiastic reception and by March 1947, 11,000 units had been shipped. A Mail-A-Voice system cost about $40. This moderate cost, and the ability to file the disks like paper, made the product appealing to the business world as a dictating machine. However, the Mail-A-Voice was designed primarily for home entertainment use; its construction was not robust enough to withstand the heavy usage encountered in the business environment. When Brush attempted to secure this more demanding market, many problems with reliability ensued, exacerbated by poor quality control and unsatisfactory interchangeability from one recorder to another.

In 1947 Brush brought out a redesigned version, which incorporated modifications that improved the device's suitability for the business markets and, in 1949, funded a project to design a completely new machine. However, after reviewing the problems with the product and increasing competition, Brush decided to direct the company efforts to other projects, and, the Mail-A-Voice was terminated in June 1950. Brush licensed the British company, Thermionic to make a redesigned version of the Mail-A-Voice. This machine, which was called the Recordon, was brought out as a dictating machine and was manufactured from 1948 until 1956.

The Brush Soundmirror

At approximately the same time the Mail-A-Voice was being designed, Brush began a parallel program to design an open-reel tape recorder that was to become known as the BK-401 Soundmirror (Fig. 6–2). The development of the Soundmirror is of particular historical significance, not only because it was the first commercial coated-tape recorder to be manufactured in the United States after the war, but because it embodied many design concepts in tape drive mechanisms, electronic equalization circuits, magnetic ring heads, and ac bias that many postwar consumer recorders built on. First demonstrated in January 1946 at the annual meeting of the Institute of Radio Engineers (IRE) in New York City, the Soundmirror was released in limited quantities in early 1947.

To reduce engineering time and speed the recorder to the marketplace, the Brush engineers used standard components where possible. The tape reels, for example, were 8 mm motion picture reels, and the drive motors were of the same type used on phonograph turntables. The tape drive components were mounted on a mounting plate that was stamped from light gauge sheet metal. As a further cost

Figure 6–2   The Brush BK-401 Soundmirror tape recorder. [*Author's collection.*]

reduction measure, a single head was used for recording and reproducing. A separate erase head was placed about an inch away from the record/play head. During record and reproduce, head-to-tape contact was maintained by felt pressure pads, which were automatically withdrawn when the recorder was stopped to allow the tape to be removed or threaded. The Soundmirror was designed to accommodate 7-inch-diameter reels that gave approximately 30 minutes of recording time at the tape speed of 7.5 in./s and produced a frequency response from 100 to 5000 Hz with a total signal amplitude variation of 6 dB. The dynamic range (the difference between the maximum sound output reproduced at a given distortion figure and the noise level) was 35 dB on 0.25-inch tape when recording at a level giving 5% distortion.

The BK-401 Soundmirror used three separate motors to drive the capstan, the take-up reel, and the supply reel. The Soundmirror and other early consumer recorders used a rubber-coated capstan to drive the tape by friction. However, the introduction of lubricated tape in 1949 necessitated a more positive drive method. A rubber-coated pressure roller that pressed the tape against a steel capstan shaft was the solution, and the majority of consumer and professional tape recorder manufacturers used this technique.

A consumer market for the Soundmirror failed to materialize because of several factors. First, the unit cost of $250 represented a substantial expenditure at that time. Second, although the performance of wire recorders was inferior to tape, their average price of $150 was appealing to the casual user. Third, the lack of prerecorded music made competition with records difficult. The major market area in 1946 proved to be among professional and semiprofessional users like musicians, schools, and individual specialists who were looking for high performance and an easy-to-edit recording medium. The U.S. broadcasting industry also started to use the Soundmirror to investigate the potential advantage of magnetic tape to record local programs and as a time-delay device. Brush sold approximately 2500 Soundmirrors in the first three months after its release in 1946. Only 28% of the total sold by 1949 were for home entertainment.

With the BK-401 Soundmirror, Brush duplicated the marketing mistakes it had made with the Mail-A-Voice recorder: both products were over priced for the target market and underdesigned for the more demanding applications for which an eager customer base existed. Under the more strenuous environments, the light-duty components failed frequently. In addition, the Brush Soundmirror was plagued with quality control problems. Brush attempted to correct the quality control issues and redesign out some of the weaknesses, offering 11 different models of the Soundmirror between 1946 and 1950. Ultimately, the reliability problems proved to be insuperable, and thus, with mounting competition, the production of the Soundmirror was stopped in 1952.

Brush never marketed a complete recorder again. However, after abandoning the tape recorder market, Brush remained active in the magnetic recording business and, under the guidance of Otto Kornei, became a major OEM supplier of high-quality custom audio and instrumentation recording heads.

The Expanding Consumer Market

Times had changed since Brush introduced the first consumer recorder to a limited market, where the majority of applications were professional. Advances in magnetic tapes and heads made it possible to reduce the tape speed while retaining a frequency response necessary to produce high-quality sound. The slower tape speed plus a reduction in the tape thickness meant that longer playing times could be achieved. These technological advances, along with reduction in cost due to mass production techniques and the growing interest in high-fidelity music, made magnetic recording appealing to the consumer.

In 1948, about a year after the Brush Soundmirror had entered the market, Eicor introduced the Model 115 half-track recorder, which was similar in design to the BK-401 with a frequency response from 80 to 7500 Hz at 7.5 in./s. With the half-track format, the tape was recorded on the upper half only; the reel was then turned over and the lower half was recorded. The half-track format provided twice the recording time on the same diameter reel.

The Revere Camera Company helped to expand this growing home market by pioneering a half-track recorder with a tape speed of only 3.75 in./s and upper frequency response of 7500 Hz to give the consumer an extended recording time of one hour on a 7-inch reel. Ampex, too, made its bid for a share of this consumer market by offering a series of both mono and stereo recorders. These machines were more expensive than the average consumer recorder of that day; however, they offered high-quality performance and were often used in less demanding professional work. By 1955 the number of U.S. companies eager to capitalize on the expanding consumer market had grown to 25, which included Pentron, Webster-Chicago, Wollensak, Bell Sound, Sears Roebuck, and Berlant Concertone.

This expansion was taking place overseas as well. The British company Wright and Weaire (an old radio component manufacturer) introduced a two-speed (7.5 and 3.75 in./s), half-track, 0.25-inch tape recorder in 1949. The Ferrograph recorder, also made by Wright and Weaire, was a high-quality machine and found use in both high-fidelity home installations and semiprofessional applications. At the same time, the German companies AEG Telefunken, Grundig, and BASF were beginning to manufacture magnetic recording products again. Grundig introduced several tape recorders as well as special tape dictating machines. Tandberg, a Swedish company, introduced a line of high-quality recorders starting with a three-speed machine that recorded on one channel and reproduced on two. This product line continued to advance, and Tandberg became a leading manufacturer of high-performance recorders.

Sony Corporation began making magnetic tape in Japan in 1948. A year later, the Sony group, under the guidance of Akio Morita and Masaru Ibuka, made a prototype recorder and, in 1950, the first Sony audio tape recorder was marketed in Japan. The machine was large, weighed over 100 pounds, and sold for $400.

SEMIPROFESSIONAL RECORDERS

In 1946, almost parallel to the Ampex Model 200 development (a story covered later in this chapter), a start-up company called Magnecord Corporation was formed by five employees from Armour Research and some investors. This venture had the full sanction of the Armour management, since the older firm, which had failed to interest its licensees in manufacturing recorders for the broadcast market, saw this as a vehicle to further promote its interests. The first product, a wire recorder built on Marvin Camras's work, was not successful for reasons of high cost and poor reliability. With the introduction of 3M tape in 1947, and reports on the German Magnetophon, Magnecord decided to shift its product direction from wire to tape. Magnecord's first tape recorder, an endless-loop machine (Audioad) offered to the market in late 1947, was designed to play recorded messages for sales promotion. High cost and design complexities doomed this machine in the marketplace.

After these two unsuccessful attempts to develop magnetic recording products, R. J. Tinkham and J. S. Boyer, the Magnecord executives, discussed their plans to build a small portable recorder for the broadcast market with Bell Labs. Following the Bell Labs consultation in February 1948, they were convinced about the future of tape. Thus the PT-6 (Fig. 6–3), a recorder specifically aimed at the broadcast market, was designed and built; it was first demonstrated at the National Association of Broadcasters (NAB) show in May 1948. By the end of the show Magnecord had sold 70 machines, and 100 more were sold in the month that followed. A network of manufacturer's representatives and dealerships was established throughout the country to sell and service the recorders. The company grew rapidly and became the dominant player in the low-cost broadcast market during the 1950s.

A significant factor that contributed to the popularity of the Magnecord PT-6 was the price: $750 compared with the $5000 Ampex Model 200. In addition, the PT-6 was much smaller and more compact than the Ampex recorder, and its sound quality was acceptable for most broadcast purposes. The recorder was composed of two separate parts, the tape drive mechanism and the record/reproduce amplifier, each measuring approximately $18 \times 8 \times 16$ inches. These units could either be installed in portable carrying cases for on-site recording or mounted in an equipment rack for studio use. The tape drive mechanism was designed to use 0.25-inch magnetic tape on 7-inch-diameter reels. As with the

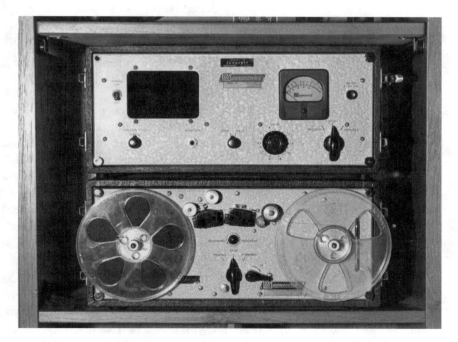

Figure 6–3 The Magnecord PT-6 tape recorder. [*Author's collection.*]

Soundmirror, the same head was used for record and reproduce, and an erase head was located before the record/reproduce head. By physically changing the size of the capstan and the accompanying rubber-coated pressure roller, the tape could be transported at two different speeds to meet different tape consumption and frequency response requirements. The frequency response was from 50 to 7500 Hz at the low speed (7.5 in./s), and 50 to 15,000 Hz at the higher speed (15 in./s). The low and high speeds provided recording times of 30 and 15 minutes, respectively. The magnetic heads used on the PT-6 were composed of a front core section with a nonmagnetic gap, which was inserted into a rear coil assembly. When the front section became worn, it could be removed from the coil assembly and replaced. This unique design saved the cost of replacing the entire head.

The Magnecord tape drive used a motor on the supply reel for rewinding the tape. A synchronous main drive motor was used to drive the capstan and to supply power to the take-up reel through a rubber-coated wheel mechanism. Instead of maintaining head-to-tape contact by means of pressure pads, which are subject to wear and require frequent cleaning, a mechanical friction mechanism attached to the rear of the rewind motor was used to apply a small drag force to the supply reel. Tape contact was maintained with the recording heads by the tape tension that, in effect, stretched the tape over the heads. Maintaining head contact by developing back tension on the tape was the method most widely used by professional tape recorders.

The recorder was used by General Motors to analyze automobile noise; however, the recordings lacked spatial perspective. In 1949 General Motors asked Magnecord if it would be possible to make a stereo recorder as a potential solution to the perspective problem. Magnecord modified the recorder by installing two record/reproduce heads about 1.5 inches apart, along with an additional record/reproduce amplifier, to make one of the first stereo recorders. This recorder was demonstrated at the 1949 New York Audio Fair, but it was not widely used because there was no stereo broadcasting at that time.

Although the Magnecord was more reliable than the Brush Soundmirror, problems were encountered with the rubber-coated wheel capstan drive system. In 1953 another model was offered to the market with a newly designed nonslip capstan drive mechanism. Although providing better reliability, this was a larger recorder and lost much of the appeal of the earlier, more compact units. Increased competition from Ampex, which began to offer cost reductions, better performance, and more reliability, was diminishing Magnecord's market share. The company made an attempt to capture a share of the high-end broadcast and studio mastering market by introducing the M-90 recorder in 1955. This machine had improved recording heads and electronics and better performance than the PT-6. However, by that time Ampex had gained a substantial foothold in this market and was offering a recorder with better performance than the Magnecord M-90 recorder, as well as operational advantages. Even though Magnecord had a significant share of the broadcast market at one time, the company lost its command and never regained that position. Magnecord was purchased by Midwestern Instruments Corporation in 1964 and in 1970 became a part of the Telex Corporation.

AMPEX PROFESSIONAL RECORDERS

The Origins of Ampex

About the same time that Brush was developing the Soundmirror, the German Magnetophon tape recorder was being introduced in the United States by Jack Mullin. During the war, Mullin had served in the Royal Air Force and, while listening to German music broadcasts, he became curious because the radio music had a "live" quality. Later, during an assignment in the U.S. Army Signal Corps in Germany, he discovered that the music had been broadcast from an AEG Magnetophon tape recorder, a K4 model using ac bias. He brought two Magnetophons back to Paris, where he was stationed, for study by the Signal Corps. He subsequently acquired two additional recorders along with German magnetic tape for himself. Mullin rejoined his prewar employer, the W. A. Palmer Studios, in early 1946 and began to demonstrate the Magnetophon at engineering meetings, and the company used the recorders in its film production work.

Harold Lindsay, who was an engineer with the Dalmo Victor Company, and an audio enthusiast, attended one of Mullin's Magnetophon demonstrations at a meeting of the IRE in San Francisco in May 1946. Four months later, the founder of the Ampex Corporation, Alexander M. Poniatoff, hired Lindsay as a part-time consultant to help find potential new products. At that time, Ampex manufactured precision motors and generators that were used in airborne radar units made by Dalmo Victor, under contract from Sperry Gyroscope and the U.S. Navy.

Ampex was founded in November 1944 in San Carlos, California, by Poniatoff, a Russian-born electrical engineer, who had emigrated to the United States in 1927. The company's name was said to be derived from Poniatoff's initials plus "ex" for "excellence." As Ampex's wartime contracts were coming to an end, Poniatoff was looking for products that could serve as the basis for a postwar business. Lindsay told his new employer about the Magnetophon, as demonstrated by Mullin, and suggested that a broadcast tape recorder based on the German product might be a suitable product for Ampex. Poniatoff expressed interest and, in October 1946, Mullin was contacted to arrange a demonstration for the Ampex chief. Conveniently, a demonstration of the Magnetophon was scheduled at the Society of Motion Picture Engineers in Los Angeles, and Poniatoff attended. This demonstration convinced Poniatoff that Ampex's postwar product should be a magnetic tape recorder. Mullin, who was a consultant to another company in addition to his work with Palmer, indicated that he would assist Ampex where such work would not conflict with his agreement with his consultancy client, Rangertone.

Rangertone, Inc., a small east coast company that manufactured the first electric organs, was owned by Richard Ranger, who had headed the electronics section of the U.S. postwar intelligence service, with the rank of colonel. Ranger, like Mullin, had been impressed with the performance of the Magnetophon and after he left the military in mid-1946, made plans to manufacture a copy of the Magnetophon. Subsequently, Rangertone manufactured magnetic recorders for sound synchronization for movie film and other specialized recorders.

Lindsay became a full-time employee of Ampex, and was made responsible for the development of a professional audio tape recorder based on the Magnetophon. He and Myron Stoloroff, an electrical engineer who designed the electric motors for Ampex, began the development work in December 1946.

The Development of the Model 200

The development of the Ampex Model 200 recorder was significant because it was the first successful U.S. implementation of the German Magnetophon technology and was key to the acceptance of magnetic recording as the new medium for the broadcast and recording industries. As a result of the Brush Soundmirror development, many of the professional users in the United States recognized the potential of the tape recorder as a production tool, but the Brush recorder did not perform to their standards. Ampex had the benefit of Jack Mullin, who had experience in the radio and film production industry and offered guidance and advice to the Ampex engineers. As a result of this expertise, and building on the Magnetophon technology, the Ampex Model 200 was the first U.S. tape recorder designed to meet the stringent requirements of the professional user.

The development of the recorder began with the heads because it was realized that they were key to the performance of the recorder. Heads require very special fabrication technology and the ability to maintain exceedingly small dimensions. The Ampex engineers had no prior knowledge in this area but, in the absence of suppliers of heads at that time (Brush did not then offer heads for sale), Ampex had to master the critical fabrication processes before recorder development could proceed. The Schüller ring head, used in the Magnetophon, was the seed for the Ampex engineers' head development work. They also made use of their experience in electric motor design and fabrication along with their good engineering sense. In the spring of 1947, after many months of "learning how," Ampex developed heads that outperformed the Magnetophon head. Ampex now had the technology to manufacture its own heads and could move on the rest of the recorder design.

In contrast to the Brush Soundmirror and Magnecord products, the Model 200 was a massive, precision-constructed machine with physical measurements of 26 × 38 × 40 inches. The tape drive mechanism layout followed the straight-line tape path, open-reel design of the Magnetophon (Fig. 6–4). Both the take-up and supply reels were powered by heavy-duty induction motors designed for continuous use. The capstan shaft was directly driven by a separate synchronous motor through a flexible coupling, which provided a reliable, positive drive mechanism with good speed regulation. Head-to-tape contact was accomplished

Figure 6–4   Harold Lindsay with the Ampex Model 200 tape recorder. [*Courtesy of Ampex Archives.*]

by applying a small voltage to the supply reel motor during the record and re-produce modes, thus producing a torque in the reverse direction to the tape mo-tion. The tape was tensioned by the forward motion of the tape pulling against the reel motor torque. Most professional recorders to follow the Model 200 used this technique. The use of three separate motors enabled the tape motion functions to be controlled electrically with push buttons, which offered greater speed and reliability than mechanical linkages. The all-electric system also al-lowed two more features important to the professional user: remote control and automation.

Because of the limitations of the early German magnetic tapes, the Magne-tophon required a relatively high tape velocity of 0.76 m/s (30 in./s) to obtain the necessary dynamic range and frequency response for high-quality audio repro-duction. Although many improvements were being made to magnetic tape which would have permitted a somewhat lower tape velocity, Ampex, in the interest of interchangeability with the Magnetophon, adopted the same speed of 30 in./s. With this tape speed and 0.25-inch tape recorded across the full width, the Model 200 achieved a dynamic range of 60 dB at 5% distortion and a frequency re-sponse of 30 to 15,000 Hz with a 2 dB maximum amplitude variation. By using a 14-inch-diameter reel, a recording time of 35 minutes could be obtained, which made it possible to record a full 30-minute program with some safety margin. Like the Magnetophon, the Ampex machine was designed to use single-flange reels on which the tape was wound tightly around the center hub so that the fric-tion between the adjacent layers would keep the tape in place. While the Ampex recorder would accept double-flange reels, single-flange reels were commonly used on early professional recorders.

Separate record and reproduce heads facilitated simultaneous reproducing while recording and were housed, along with the erase head, in a modular plug-in assembly similar to the Magnetophon. The individual heads were heavily shielded to reduce noise and stray signal pickup, and a movable front gate pro-vided access to the head for cleaning and editing. The heads were connected to separate amplifiers, which contained the record and reproduce equalization re-quired to achieve a flat response from the recorder. Since the amount of equal-ization required is a function of the tape characteristics, the equalization circuits of the Model 200 were designed to be easily altered to keep pace with the rapid changes taking place in magnetic tape development. The entire signal system was mounted on an isolation mount to reduce the effects of microphonic noise from the vacuum tubes. The ac bias was derived from a high-frequency oscilla-tor, which was connected directly to the record head. To provide a low-noise ac erasure of the tape, the erase head was also driven by the same bias oscillator. The bias frequency was set at 100 kHz. To prevent interference effects (beat frequencies) between the audio and bias signals, professional audio recorders typically used a bias frequency of four to six times the highest recorded audio frequency.

The Introduction of the Model 200

All during the development phase, Ampex had to struggle to finance the Model 200. It was very fortuitous for Ampex that Bing Crosby and his sound engineers were investigating methods other than phonograph transcription to record the popular singer's radio show.

Crosby disliked doing live radio performances because of the demands on his time and the highly structured format. In 1944 he quit radio because he was unable to convince NBC to allow him to prerecord his shows. Then a dramatic change in the network structure took place in 1946 as the result of a government antitrust action that resulted in the sale of one of NBC's two networks. The American Broadcasting Company was the outgrowth of NBC's divestiture. ABC lured Crosby back into radio by allowing him to prerecord his programs using 16-inch phonograph transcription technology, but he was not satisfied with the relatively low quality sound.

In mid-1947 Mullin demonstrated the Magnetophon recorder to Crosby and his engineers as a possible means of recording the Crosby shows. The entertainer and his crew were impressed but expressed concern that there were only two machines, with no replacement parts. At that point, Mullin told them of Rangertone's plans to make a copy of the Magnetophon recorder that could be a possible backup.

In August 1947 the Crosby engineers made a test comparison between the Magnetophon and the Rangertone copy and found the U.S. model unacceptable. They proceeded to record some Crosby shows on the Magnetophon. Because the Rangertone recorder development had not progressed as anticipated, Mullin informed Crosby of Ampex's recording efforts. In September 1947, when the Ampex Model 200 was completed, a demonstration of the recorder was made to Crosby and his engineers. Crosby ordered 20 recorders.

Ampex had been preoccupied with technical development, and no plans to market the machine had been made. As a part of the Crosby deal, Ampex agreed to make Bing Crosby Enterprises the distributor for the recorder in the western United States. Crosby also made an advance payment, which gave Ampex the much needed capital to expand its facilities in order to produce the recorders. Two Model 200 recorders were delivered in April 1948 for recording the Crosby shows; the remaining 18 machines went to ABC's Chicago studios and were used to time-delay broadcasts. A short time later, Decca Records in the United States purchased Ampex recorders to record master tapes from which phonograph records were made. NBC also bought Ampex machines to replace its Brush Soundmirrors. The Ampex recorder venture was paying off.

By the end of 1948, Ampex had sold 112 machines. The Model 200 had gained a reputation for quality and reliability, but its high cost placed it out of reach of many potential users. Under competitive pressure from the PT-6 Magnecord recorder, just entering the market at a price of $750, Ampex lowered the initial selling price of $5000 to $3000. The Model 200 was somewhat overdesigned and expensive to produce, but the Ampex founders made few cost reductions because they did not want to risk compromising quality and reliability. However, Ampex realized that to stay competitive and build on a good reputation, a lower cost design with comparable reliability and quality was needed.

Thus in November 1948 work was started on a new recorder, which would later be called the Model 300. The improvements in magnetic tape and heads made since the introduction of the Model 200 allowed the Ampex engineers to cut the tape speed in half while retaining the same frequency response and dynamic range. The speed reduction to 15 in./s and thinner tape allowed the tape reels to be reduced from 14-inch to 10.5-inch diameter and retain the same recording time. The machine was designed to use the new tape reel sanctioned by the National Association of Broadcasters (NAB reel) instead of the open single-flange reel of the Model 200. The Model 300 tape drive mechanism was essentially unchanged from the Model 200, but a new head assembly was designed which provided easier access to the tape for editing. These design improvements resulted in a recorder that was less costly to make but retained the quality and reliability of the higher priced model. Size reductions were also made to the signal system electronics and other components of the machine, and as a result the Model 300 was much smaller than the Model 200. The first production of 50 units was released in May 1948. Well received by the broadcast and recording industries, the machines subsequently became the standard for record mastering.

The Model 300 transport design was a building block for many future Ampex products. Modified versions were used later as tape duplicating machines and as early data recorders used by NASA and others. During the product life, which extended into the 1970s, over 32,000 of various versions of the Model 300 were sold.

Ampex followed the Model 300 with its first portable recorder, the Model 400, introduced in 1949. In 1952 Ampex began manufacturing the Model 350. This highly reliable and easy-to-maintain recorder became one of Ampex's most popular products. Smaller and less expensive than the Model 300, the Model 350 was aimed at the high-end broadcast market. In 1954 Ampex introduced the Model 600, a small half-track portable professional recorder, measuring $18 \times 12 \times 9$ inches. The Model 600 was used extensively as a broadcast "on the scene" recorder.

Ampex's early success was primarily due to Bing Crosby's desire to prerecord his radio programs, and the ABC network's willingness to break with the traditional network structure by accommodating Crosby in this respect. Thus the opportunity arose for magnetic recorders to be used as a production tool to prere-cord programs. The Model 200's timely development and its high performance and reliability put Ampex in a position to take advantage of the opportunity that Crosby's patronage afforded. Early on Ampex earned a reputation for producing high-quality products with a high degree of reliability. In time, user confidence grew, and the skeptical networks became convinced that magnetic tape was a reli-able and dependable tool, resulting in a complete change in the philosophy of net-works toward magnetic recording.

Ampex's continued success was built on maintaining its reputation for high performance and reliability. Through heavy investment in research and develop-ment, Ampex continued to demonstrate creative engineering abilities through the development of new and more advanced products. Ampex's competitive strategy was to produce quality, state-of-the-art, high-end products that would command a premium price. Moreover, Ampex from its early association with Bing Crosby maintained close ties with the entertainment industry. This relationship gave Ampex valuable marketing insights that facilitated the development of products that were directed toward this industry.* Ampex, in turn, was looked on as a re-source that was responsive to entertainment industry needs. By the mid-1950s, its reputation for high-quality products well established, Ampex dominated the high-end audio market.

OTHER PROFESSIONAL RECORDERS

As the Ampex Model 200 was nearing completion, EMI in the United Kingdom produced that country's first professional studio recorder. EMI's model BTR1, which appeared in November 1947, was also built on the Magnetophon design. Large numbers were sold to the BBC, and the BTR1 progressively replaced the disk-cutting lathes at EMI's Abbey Road Studios for master recording. The tape speed was initially set at 30 in./s, to obtain reasonably high quality from the early European tapes. When EMI began to manufacture improved magnetic tape, the BTR1 was changed to a two-speed machine operating at both 15 and 30 in./s.

As magnetic recording became established in the broadcast and recording industry, well-established U.S. manufacturers of studio and broadcasting equip-ment like Presto, Fairchild, and RCA, started offering magnetic recorders to com-plement existing product lines. Stancil-Hoffman manufactured professional audio recorders that were used by the Armed Forces Radio Service for continuous recording and reproducing.

---

*This relationship deteriorated rapidly when Bing Crosby Enterprises began manufacturing instrumenta-tion recorders in direct competition with Ampex—see Chapter 20.

CONCLUSION

The postwar period was an era of refining the magnetic recording technology that had been created in early years in Germany and, to a lesser extent, in the United States. Magnetic tape oxides were developed with better oriented, smaller particles, along with improved coating and surface finishing techniques. These advancements provided more intimate contact between the head and tape, making it possible both to record and reproduce higher frequencies and to improve the dynamic range. At the same time, new head materials with better magnetic properties, and improved fabrication techniques, resulted in more efficient heads with less high-frequency loss, thereby improving the signal-to-noise ratio. The recognition by radio networks and local broadcast stations of the performance and economic advantages offered by magnetic tape brought about changes in attitudes toward live programming. By the mid-1950s, magnetic audio recording had completely revolutionized the record and broadcasting industry. All records were mastered on tape, and radio broadcasters were exclusively using tape as a time-delay and programming tool.

References

Begun, S. J., *Magnetic Recording*, Murray Hill Books, New York, 1949.

Camras, M., *Magnetic Tape Recording*, Van Nostrand Reinhold, New York, 1985.

Camras, M., *Magnetic Recording Handbook,* Van Nostrand Reinhold, New York, 1988.

Clark, M. H., "The Magnetic Recording Industry, 1878–1960: An International Study in Business and Technological History," Doctoral dissertation, University of Delaware, 1992.

Lane, B., "75 Years of Magnetic Recording," *Wireless World,* March, April, May, and June 1975.

Lindsay, H., "Magnetic Recording, Part 1," *dB Magazine,* 38–44, December 1977.

Lindsay, H., "Magnetic Recording, Part 2," *dB Magazine,* 40–44, January 1978.

Mallinson, J. C., *The Foundations of Magnetic Recording,* Academic Press, San Diego, CA, 1993.

Mee, C. D., and Daniel, E. D., *Magnetic Storage Handbook,* 2nd ed., McGraw-Hill, New York, and IEEE Press, Piscataway, NJ, 1996.

Morton, D. L., "The Rusty Ribbon: John Herbert Orr and the Making of the Magnetic Recording Industry," *Business History Review,* **67**, 589–622, 1993.

Mullin, J. T., "Creating the Craft of Tape Recording," *High Fidelity Magazine,* 62–67, April 1976.

Van Prang, P., *Evolution of the Audio Recorder,* EC Designs, Inc., Waukesha, WI, 1997.

7 Product Diversification

Mark H. Clark

By the mid-1950s, magnetic recording had become a firmly established technology with extensive application in radio broadcasting, music production, and high-end consumer audio. Over the next 30 years, manufacturers continued to develop more and more sophisticated open-reel analog recording machines and tape, and to supplement these improvements by introducing electronic means of noise reduction. Magnetic recording became universally accepted as the means of making master recordings for subsequent duplication and distribution of high-quality music programs to the public, primarily in the form of phonograph records. This practice continues today, although digital audio magnetic recording began to replace the analog format in the late 1970s.

An important development in professional audio recording between 1955 and 1975 was the introduction of master recorders capable of recording on a large number of tracks. Although a minor innovation in terms of technology, multitrack recording had major practical and artistic implications. In particular, the resulting flexibility of editing allowed the creation in the studio of types of music that previously were difficult, if not impossible, to achieve.

In the consumer market, the primary focus was on developing products that were easier to use. In the decade after the war, open-reel tape recorders became popular with audio enthusiasts, who enjoyed the novelty of being able to record their own voices, or favorite programs from the radio or phonograph records. A market also developed in industry and education, where consumer equipment was adapted for semiprofessional use. As the price of recorders declined and as features like stereo recording became available, demand increased, and the term "tape recorder" came into common usage.

However, the availability of prerecorded music on tape was very limited. Phonograph records continued to be the principal means for the dissemination of music. The sound quality of a phonograph LP record was excellent—stereo LPs were available after 1957—and the operation of a phonograph was simpler than threading tape in an open-reel recorder. Manufacturers came to believe that if a method could be found to avoid the threading process, tape recording could become a more viable means of providing the consumer with prerecorded programs, particularly in automobiles, where phonographs were unsuitable.

In pursuit of this goal, several companies sought to develop new formats that would make recorders easier to operate. These efforts took two paths: the cartridge and the cassette. In the cartridge, a single tape reel was enclosed in a shell, with provision for either automatically feeding the end of the tape into the transport, or forming the tape into an endless loop running from the inside to the outside of the reel. In the cassette, feed and take-up reels for the tape were housed side by side in a shell that could be easily inserted into a tape transport.

These two parallel developments—high-quality recording equipment for professionals, easier-to-use equipment for consumers—are the primary focus of this chapter. At the same time, it covers the various incremental developments in recording technology that allowed considerable increases in sound quality at lower cost during this period.

PROFESSIONAL AUDIO RECORDING

High-Quality Recorders for Broadcast Use

By the mid-1950s, professional tape recorders, such as the those made by Ampex in the United States and EMI, Grundig, and Studer in Europe, were established as a reliable and accepted part of the professional audio recording industry. Audio quality was more than adequate to meet the needs of AM and FM radio; and the convenience and economy of use proved enormously beneficial to the broadcasting industry. The use of phonographic disk recorders by broadcasting companies rapidly became a thing of the past. Magnetic tape recorders changed the way in which live programs could be rebroadcast in different time zones, facilitated the insertion of commercials, and formed the basis of preparing news programs. The latter included the use of portable battery-operated recorders in the field, such as the Swiss-made Nagra.

Master Recorders for Studio Use

Also during the mid-1950s, tape recorders increasingly found a place in the studios of music companies, which had previously used phonographic recording exclusively. The ease with which tape recordings could be edited was a particularly attractive feature. By cutting and joining tape together, the recording engineer could edit out mistakes or combine parts of two or more performances to create a single more polished version.

Initially, the signal-to-noise ratio of the professional recorders was marginally adequate to meet the requirements for making the highest quality master recordings, particularly if significant rerecording took place. This problem was addressed by Ray Dolby, famous for his part in developing the Ampex Quadruplex video recorder. More important in the present context, Dolby was a classical music enthusiast who wanted to improve the quality of studio recordings. At his laboratory, then based in a suburb of London, he developed the first of his noise reduction systems, the A type, in the mid-1960s. The key to his invention was the compander concept, involving precompression of the signal before recording followed by complementary postexpansion after recording, with the object of leaving the signal unaltered while suppressing background noise. Previous attempts to use this concept had failed because they produced unpleasant side effects—the systems could be heard "breathing." Dolby avoided such difficulties by adjusting the compander to act differently in different parts of the dynamic range and the frequency spectrum. The result was an improvement in signal-to-noise ratio of 10 dB, with virtually no noticeable degradation of signal quality. More recent Dolby systems, designated SR (spectral recording), use sliding filters controlled by the spectral content of the signal.

Other noise reduction systems were developed around the same time (e.g., DBX, Telcon), but the Dolby system prevailed and eventually became the worldwide standard for preparing and exchanging mastering tapes. The Dolby units were not inexpensive—initially about $2000—but cost was of secondary importance to achieving the requisite level of sound quality for mastering purposes.

Multitrack Recorders for Popular Music

For much popular music, the emphasis was not so much on using magnetic tape mastering to achieve the ultimate in sound quality, but on using tape to provide economy of operation and flexibility in editing. The primary result of magnetic tape usage was a widening of the scope and character of the music handled by the recording industry. By the late 1950s, small regional recording studios could afford equipment that allowed them to compete with the majors. This led to a significant change in the nature of popular music since, simultaneously, a major change was occurring in the radio industry.

In the postwar period, television took over the traditional programming of radio—soap operas, children's shows, and prime-time dramas and comedies. Increasingly, radio stations played music, and only music. As a result of the change in programming, radio audiences, particularly teenagers, were able to listen to music that would have been unavailable to them only a few years before. By the early 1960s the major record labels had adapted to the public's changing taste in music, an accommodation that ushered in the era of rock and roll as the dominant musical style in the United States. Rock and roll was youth-oriented, and marketed as part of the larger culture of youthful rebellion against convention.

Prior to the mid-1960s, a professional studio recording essentially sought to duplicate the feel of a live performance for popular as well as classical music. Musicians performed together at the same time, and a perfect recording required the simultaneous achievement of perfection by all the musicians, and by the mixing engineer. Mixing consoles had a limited number of inputs, and limited or no equalization; commercially available recorders had only three or four tracks.

After the mid-1960s, musicians, particularly rock musicians, increasingly demanded multitrack equipment in the studio for recording. Multitrack equipment allowed a musical recording to be assembled by recording one track at a time. The sound from one instrument or voice was recorded on a single track, and each such segment could be electronically manipulated with equalizations, echo, or other techniques, without affecting the performances on any other track. If the performance was unsatisfactory, the track could be erased and rerecorded without affecting what was on the acceptable tracks. When all the parts of the performance were satisfactory, the contents of the various tracks were combined to make the final master recording that would be released to the public. Multitrack recording also allowed artists to do things impossible outside the studio. A musician could play a duet with himself; a singer could harmonize with her own voice; a determined and versatile performer could record an entire symphony by playing each instrument in turn. The possibilities were endless, and very attractive to those with a creative bent.

Increasing use of these techniques not only put pressure on manufacturers to develop new multitrack equipment, but provided considerable opportunity for new companies to enter the market for professional recorders, since the state of the art was changing constantly. The first wave of multitrack machines came from American companies. Ampex and Scully, the two firms that dominated the professional market in the early 1960s, were joined by the tape maker 3M in 1966. By 1968 8-track machines had become standard, with 16- and 24- track machines introduced shortly thereafter (Fig. 7–1). Several other firms entered the market at this point, most notably the Swiss company Studer. Studer eventually developed a popular 40-track machine that used 2-inch-wide tape and, by the early 1980s, studios with 48-track capability were not uncommon. The constant expansion of the number of tracks created a related industry—mixing consoles that automated fading, to make it possible for the recording engineer to actually use all those tracks.

During the same period, improvements in tape and head quality allowed other firms to introduce less expensive multitrack machines suitable for smaller studios. In 1969, for example, TEAC marketed a 4-track synchronized machine that used 0.25-inch tape, generally recognized as the first affordable multitrack machine. Other small firms such as Dokoder, Otari, and Fostex also entered this market in the early 1970s, making home studios possible.

Figure 7–1   An Ampex 1100 studio mastering recorder using 2-inch tape and
24 tracks. [*Courtesy of Ampex Archives.*]

In the mid-1970s the first digital recording multitrack studio recorders were introduced, though the lack of common standards for encoding delayed their widespread use until the 1980s. Eventually, however, the ability of digital machines to make copies without generational loss of sound quality led more than half the studios to make the switch. This change is discussed in Chapter 8.

CONSUMER AUDIO RECORDING

Open-Reel Recorders

In the 1950s manufacturers of less expensive open-reel recorders established a significant base in the consumer market by selling to people who wanted a means of making their own recordings. In the United States, Brush and Armour entered the field early, but their recorders were expensive and poorly designed.

In Japan, at a small start-up company called Sony, the founders, Akio Morita and Masaru Ibuka, were looking for their first major product. They produced the first Japanese-made open-reel tape recorder. By 1960, the open-reel business was booming, with participation by many companies from the United States, Europe, the United Kingdom, and Japan (Fig. 7–2).

However, the phonograph record player still reigned supreme for prerecorded music—a situation that would not change for another 20 years. The owners of most copyrighted music were reluctant to make their programs available in the form of prerecorded tapes for use in the home, where they would compete directly with the copyright owners' bread-and-butter product, the phonograph record. Tape scored one early success in being the first to introduce stereo recording to the public; but this advantage vanished when stereo LP records were developed by Westrex and marketed by CBS and RCA in the late 1950s. Prerecorded open-reel tapes were also very expensive—$12.50 for a 30-minute stereo program versus $4 for an LP.

Three other factors were against the use of tape as a player of prerecorded programs. First, tape players were more expensive than phonographs of comparable quality. Second, unlike a phonograph album, where the content on each side of the disk was rapidly accessible, tape players required the user to wait while the tape was fast-forwarded or rewound. Third, it was more difficult to thread a tape through the feed system and past the heads than to put a tone arm down on a phonograph disk.

Figure 7–2   An open-reel recorder of the mid-1960s, made by UST of Japan for Ampex. [*Courtesy of Ampex Archives.*]

In the belief that the last factor was the critical one, tape recorder manufacturers spent a great deal of effort during the 1950s and 1960s developing new systems that would take the place of the open-reel machine, particularly for automobile applications.

The 8-Track Cartridge

The obvious solution to the problem was to develop systems that enclosed the tape in a container that could be easily loaded and removed. The first such system that became widely popular was the cartridge, the best-known example of which was the 8-track, 0.25-inch tape system. Unlike a cassette, which essentially duplicates a open-reel recorder configuration, the 8-track cartridge contained only a single reel. The tape was an endless loop, running from the innermost layer of tape on the reel, through the transport and over the playback head, then back onto the outermost layer of the same reel. Upon being inserted into a player, the tape played back continuously; since the loop repeated endlessly, there was no need to rewind.

The first successful magnetic tape endless-loop cartridge was developed by Bernard Cousino in the early 1950s. The owner of a small audiovisual sales and service company in Toledo, Ohio, Cousino had received a contract to build a device that would endlessly repeat a short advertising message out loud. Being familiar with endless-loop film cartridges, he used 0.25-inch magnetic tape in an 8 mm film cartridge as the basis for his experiments. The tape tended to bind up, however, not so much because of friction, but because of the static electricity generated by rubbing between layers of tape. Cousino solved this problem by coating the back of the tape with colloidal graphite, which both lubricated the tape and conducted away the static electricity. The back-coated tape introduced new problems, however, such as head clogging.

In 1952 Cousino sold a cartridge under the name Audiovendor for use with point-of-sale display advertising. This cartridge was used with a standard open-reel player by placing it over one reel spindle and guiding the tape over the existing heads. Some years later, an improved two-track cartridge was produced, the Echomatic, which fully enclosed the tape and required a special machine to record and play. These machines were sold primarily to the point-of-sale and educational audiovisual market.

Cousino's work inspired others working in audio recording, including George Eash, an inventor who rented space in Cousino's building. Eash designed and developed the Fidelipack cartridge in the mid-1950s. By the late 1950s, the Fidelipack had become the dominant cartridge design in commercial applications, despite its relative complexity and proneness to sudden failure. The two major applications of the Fidelipack were in radio station announcing and automobile sound systems. Several firms designed and marketed machines specifically designed for radio station use, with heavy-duty mechanisms that included provisions for remote control and end-of-tape sensors. By the early 1960s, these machines were widely used for frequently repeated on-air messages, such as spot announcements, advertisements, and station identification. The success of the Fidelipack system led other firms to develop and market similar products for the radio broadcast industry during the 1960s and 1970s, including the Audiopac, Aristocart, Marathon, and Tapex systems.

The key person behind the development of cartridge systems for use in automobiles was Earl Muntz, an automobile dealer from southern California. Muntz first learned of the Fidelipack in the early 1960s. He sold his television business and used the proceeds to set up a new firm to market Fidelipack players for the automobile sound market. Muntz adopted the recently established stereo tape standards for his players, putting a total of four tracks (two stereo programs) on the standard 0.25-inch tape. The playback unit had two heads mounted together, which moved by means of a solenoid from one set of tracks to the other when triggered by a small piece of metal foil attached at the splice that formed the endless loop. Total playing time was 40 minutes. Although stereo was not strictly necessary, since the overall quality of auto sound systems of the time was relatively poor, Muntz used the feature to identify his products with the prestige of "hi-fi" audio equipment, which had recently come to include stereo reproduction.

The growth of Muntz's 4-track system was soon halted, however, by the arrival on the market of a new cartridge system, the 8-track. This format was the brainchild of William P. Lear, developer of the Learjet business plane. In 1963 Lear became a distributor for Muntz's Fidelipack machines, primarily so that he could install them in his Learjets. Lear soon became dissatisfied with the system, however, and the same year proceeded to develop his own version. He began by having his engineers incorporate a new head, which allowed him to double the number of tracks on the same tape width, increasing playing time to 80 minutes. The cartridge was also redesigned, reducing the number of parts and incorporating an integral pressure roller (Fig. 7–3). In 1964 Lear's aircraft company built a hundred players based on this new design. Lear distributed them to executives in the auto industry and at RCA.

As a result of Lear's efforts, the Ford Motor Company agreed to install 8-track players in its cars, beginning with the 1966 model year, and RCA Victor agreed to make its catalog of music available on 8-track cartridges. The players were manufactured by Motorola, Ford's primary electronics supplier and a pioneer in the development of car radios. Consumer response was much larger than expected. Ford installed 65,000 players in the first year, and sales continued to grow by leaps and bounds thereafter. Both General Motors and Chrysler offered 8-track players beginning in the 1967 model year. By the end of 1967, an estimated 2.4 million 8-track players were in use. The success of the auto market motivated manufacturers to offer home and portable 8-track machines beginning in

Figure 7–3   The 8-track cartridge. [*Courtesy of Ampex Archives.*]

the late 1960s. As a result, 8-track was the preeminent portable and car audio format of the early 1970s.

After 1965, manufacture of 8-track machines moved increasingly from the United States to Japan. With the exception of a few American producers of high-end equipment such as Quartron, by 1975 the makers of 8-track equipment were all Japanese. This mirrored a larger trend in the American electronics industry at the time, as consumer products increasingly were of Japanese manufacture. The sale of 8-track equipment was largely restricted to the United States; the format never did well overseas.

The format began dramatically losing market share after 1975. Momentum shifted to the competing cassette format, which had improved rapidly from its humble beginnings in the early 1960s. Most major record labels abandoned 8-track in the early 1980s as a way to simplify production and inventory in the wake of a general move to the compact disc.

The cartridge format in general, and the 8-track in particular, currently can be said to have failed as a technology. From a marketing point of view, thanks largely to its promotion by the auto industry, the 8-track cartridge was by far the most successful new audio format of the mid-1960s and early 1970s. From a technical point of view, however, the 8-track was inherently an unreliable design, subject to frequent mechanical problems, and missing the basic advantages of conventional tape machines: namely, fast forward and reverse, easy erasure, editing, and indexing. Although tolerable when the 8-track was used as a player in an automobile, these limitations were a serious impediment to the use of this format for recorders in the home.

Early Tape Cassettes

The idea of using a cassette system mirroring the configuration of an open-reel recorder has a long history. As early as 1912, the American Telegraphone Company developed a locking clamp that allowed an operator to remove and replace the reels of a wire recorder simultaneously while maintaining them in the same relative positions. In the early 1930s, the Echophone Company of Germany enclosed the reels of its wire recorder in a metal case to ease handling—it was the first commercial cassette system for magnetic recording. During the second world war, both the Brush Development Company and the Armour Research Foundation designed wire recorders for the American military that incorporated cassettes.

In the period immediately following the war, a number of firms that licensed patents from Armour made plans to market cassette systems using wire as a recording medium, but because of the rapid success of coated tape, those designs failed to find a market. Moreover, most of the early sales of tape recorders were to customers who were more concerned with performance and reliability than with ease of use. As a result, although the American military continued to buy and use cassettes for specialized audio and data recording machines, commercial cassette systems were not widely available until the late 1950s.

One other reason for the delay in developing tape cassette systems was the question of bulk. At the relatively high tape speeds of the late 1940s and early 1950s, large amounts of tape were required for recordings of even modest length. However, by the late 1950s, incremental improvements in head design, recording media, and manufacturing consistency had made it possible to record at lower speeds.

It was in this context that RCA became the first major manufacturer to introduce a commercial audio tape cassette system. RCA announced its "cartridge" (really a cassette) tape system in early 1958 and advertised it at electronics trade shows that summer, although machines did not arrive in stores until August of the following year. The RCA system used a standard 0.25-inch tape operating at 3.75 in./s to give a total recording time of 60 minutes. The cartridge was bulky, measuring roughly 5×9 inches. Despite the announced backing of two tape makers (3M and Orradio Industries) and an equipment maker (Bell Sound), and the efforts of RCA itself, the project went nowhere. RCA failed to appeal either to consumers or to the hi-fi market, since the machines produced were relatively expensive, yet included features that hi-fi enthusiasts did not want, such as a built-in amplifier. Those disadvantages, combined with the difficulty of getting dealers to stock a wide selection of prerecorded cassettes, kept sales low.

The research division of CBS took a more adventurous approach in the late 1950s, although the resulting product was embodied in a cartridge, not a cassette format. CBS not only reduced the tape speed a step further to 1⅞ in./s, but also reduced the tape width to 0.15 inch. The system used a single reel (a second reel was permanently attached to the player), incorporating an automatic threading device. Recording took place in only one direction instead of two, but the program was recorded in three-track stereo. Licensed to the Revere Corporation, which made the recorders, and 3M, which made the tape cartridges, the machines cost $450—a large sum by the standards of the time. This system also failed, since it was too expensive for the average consumer, and the hi-fi market was not impressed by its performance.

Introduction of the Philips Compact Cassette

Although the RCA and CBS systems were unsuccessful, both played roles as precursors to the development of the Philips compact cassette. Initially, Philips, already a major phonograph record producer and tape recorder manufacturer, had not set out to develop its own cassette, although it had an exploratory program under way. In fact, representatives from Philips traveled to New York City in 1960 to negotiate a license for the CBS cartridge design. But the meeting got off to a bad start, and no license was negotiated.

Back in the Netherlands, in 1961, Philips proceeded to work on the independent development of a cassette system. Philips had been making magnetic recorders since the early 1950s, so the company was already experienced with the demands of sound recording. The Philips engineers rejected the RCA design as too bulky and the CBS design as too costly and complicated. They also rejected the cartridge concept, largely on the basis of the inconvenience of having to rewind the cartridge before removing it. As it turned out, their final design combined the best features of each approach. They copied one feature of the RCA system: that was to dispense with flanged reels and use the sides of the cassette to guide the tape between naked hubs. This allowed the use of a smaller cassette, since the hubs could be located closer together without the problem of interference. They adopted the narrow 0.15-inch tape width and low 1⅞ in./s tape speed of the CBS system. The net result was a cassette package of very small size—approximately $4 \times 2.5 \times 0.5$ inch. A prototype cassette received mixed reviews in 1962: good in Europe, but poor in the United States.

The first version of the compact cassette, initially known as the Pocket-Cassette, was launched in 1963 in Berlin. Philips engineers later claimed that five considerations had driven the design of the Compact Cassette:

1. Smallest possible dimensions with a playing time of 30 minutes
2. Simple sturdy construction
3. Reliability
4. Maximum protection of the tape
5. Low energy consumption during playback and rewind

These design goals were derived from the initial purpose of the machine that would use the cassette—a compact battery-operated tape recorder for dictation purposes and, consequently, low-fidelity sound recording. The desire for simplicity, durability, and above all low energy consumption (to extend battery life) was basic to the final application. To achieve the first goal, Philips engineers made use of recently available thinner tape to reduce the size of the full tape pack. The other goals were addressed by making the cassette case as simple as possible, using molded plastic for strength and ease of manufacture, and devising an uncomplicated feed path. Friction was minimized through the use of flanged rollers as guides, rather than fixed posts, and by lining the housing with Teflon wafers (Fig. 7–4).

The initial specification called for monaural recording—two tracks, one in each direction. This was soon supplemented by a standard for stereo recording. Unlike existing stereo standards, which maximized separation of the four tracks required by two-direction recording by interleaving them, the Philips standard placed the two forward tracks and the two reverse tracks next to one another This made it possible to play stereo recordings on monaural machines without modification.

The initial version of the compact cassette recorder could record for 30 minutes in each direction (C 60 cassette). By the late 1960s, thinner tape allowed for 45 minutes each way (C 90). In the 1970s a 60-minute version was developed (C 120), but it proved less popular because the extremely thin tape was prone to jam.

Licensing Tactics

The key to the eventual worldwide success of the compact cassette was not based on its initial performance. Rather, it was the way the design was marketed, and the dramatic way in which its performance was improved during the 1970s. From the beginning, Philips pursued a strategy of licensing its design as widely as possible. According to Frederik Philips, president of the firm at the time, this policy was the brainchild of Mr. Hartong, a member of the board of management. Hartong believed that Philips should allow other manufacturers access to the

Figure 7–4 The Philips compact cassette. [*Courtesy of Philips and Ampex Archives.*]

design, turning the compact cassette into a world product. Although Philips would have only a small part of this world market, Hartong thought this share would be much larger than the business Philips would do if it tried to exploit the technology alone.

The liberal licensing policies of Philips worked with other firms as well. Despite initial plans to charge a fee, Phillips eventually decided to offer the license for free to any firm willing to produce the design. Several firms adopted the compact cassette almost immediately, including a number of Japanese electronics manufacturers. By the mid-1960s over 250,000 recorders had been sold in the United States alone, and demand for the machines continued to increase over the following decade, even with strong competition from the 8-track format. Japan became the major source of cassette recorders, supplying American electronics firms, such as RCA and GE on an OEM basis. The Japanese manufacturers were able to undercut the relatively high price of the early Philips cassette drives. Manufacturing continued in Europe, with Philips and Grundig playing major roles.

During the compact cassette's first few years, sound quality was mediocre, marred by background noise, wow and flutter, and a limited frequency range. While ideal for voice recording applications like dictation, the compact cassette was marginal for music recording. By the early 1970s, however, the compact cassette had improved so much that as well as being more reliable, it had fidelity equal to, or better than, that of 8-track. Moreover, in contrast to the 8-track, the compact cassette continued to improve in sound quality in the two decades after 1970, until it gained the stature of "high fidelity" and, at best, could come close to the quality of an LP phonograph recording.

The improvements in cassette performance occurred in two ways. First, advantage was taken of every available advance in component design and new materials, with the primary emphasis on magnetically more potent, low-noise media. Second, electronic noise reduction systems were developed to provide a further substantial improvement in signal-to-noise ratio.

Cassette Tape Improvements

Eliciting high performance from the compact cassette, with its narrow track width and low tape speed, demanded a great deal from the magnetic medium, consisting initially of conventional gamma ferric oxide coated tape. The first step toward improved performance was to develop smaller particles of gamma ferric oxide, then pack them more tightly into the coating, and finally develop techniques to produce an ultrasmooth coating surface. Combined efforts on the part of pigment and tape manufacturers, spread over a decade or more, paid off by producing significant reductions in noise and improvements in high-frequency response. A highlight of these endeavors was the development by Pfizer Corporation, in the early 1970s, of a superior iron oxide, called 2228.

Meanwhile, attractive new magnetic particles had begun to enter the picture in the 1970s. These higher coercivity particles required a substantially larger drive current on recording. This was a major impediment to their acceptance in data and video recording applications, where standards were rigid and hard to change. But the manufacturers of audio cassette recorders were willing to adapt their equipment to use the new particles by incorporating a switchable increase in bias current to the record head. Philips, which controlled the operating specifications of the cassette, went along with these changes. Such flexibility on the part of the audio machine designers and manufacturers led to the audio cassette becoming the proving ground for new magnetic coating ingredients.

Chromium dioxide made its debut in the compact cassette. Developed by Du Pont in the late 1960s, its use in cassette recorders was pioneered by Henry Kloss, president of Advent Corporation in the United States, and by Nakamichi in Japan. Memorex Corporation was the first to market chromium dioxide cassettes in substantial volume (BASF and Sony were also Du Pont licensees). Chromium dioxide was expensive, but the quantity of oxide in a cassette was so small that cost was not a major factor. Subsequently developed cobalt-modified gamma ferric oxides matched chromium dioxide in coercivity and overcame certain instability problems that had plagued earlier versions of cobalt–iron oxides.

Finally, in the 1980s, two forms of metallic media were introduced into magnetic recording via the compact cassette. The first (MP) used coatings of metal (iron) particles, offering yet another increase in coercivity and a substantial increase in the achievable level of recorded magnetization. The second (ME) replaced the conventional particulate coating with a thin evaporated (vacuum-deposited) metal (cobalt alloy) layer, providing still greater magnetization, particularly at the higher recorded frequencies. Dual-layer media were also developed in which a thin outer layer of high coercivity particles was coated over a lower coercivity underlayer.

From the late 1970s onward, major tape manufacturers in the United States (Ampex, Memorex, 3M), Europe (Agfa, BASF), and Japan (Fuji Film, Maxell, TDK) produced "premium" cassette tapes of various kinds, which they marketed vigorously through heavy television advertising and promotional efforts.

Electronic Noise Reduction

The second approach to improved cassette performance was to incorporate an electronic noise reduction system based on the "compander" concept described earlier. Initially, the Philips engineers were reluctant to incorporate this type of noise reduction system, which would necessitate changing the standards for the recording process. They argued that improvements in tape and heads would be sufficient. Around 1970, however, several compander-based systems were developed and marketed, including some by the major Japanese cassette recorder manufacturers. For noise reduction to be widely accepted, it was essential that a single standard be adopted; otherwise recordings would not be interchangeable. After some vigorous technical and marketing negotiations, the format that eventually became the standard worldwide came, once again, from Dolby Laboratories.

The Dolby consumer system, designated B type, was derived from the professional A type used in recording studios. To reduce costs to a level that was practical for a consumer item, the B type used a single-channel approach, but this restriction was offset by using a variable high-pass filter controlled by the level in a differential signal path. The effect was to boost low-level, high-frequency signals by 10 dB on record and attenuate them by the same amount on playback, for a net improvement in signal-to-noise ratio of 10 dB. Initially, the cost of the system was relatively high, adding an appreciable increase to the price of a recorder. The improvement in performance was, however, sufficient to persuade the more discriminating consumers to pay a premium. The audio pioneer Henry Kloss was instrumental in promoting early use of the B type, first in an open-reel recorder, then in his Advent recorder. In time, mass production using large-scale integrated circuit (LSI) technology reduced the price to a small fraction of the cost of a recorder, and the Dolby system was incorporated in all but the lowest price recorders.

Later versions of the Dolby noise reduction units gave larger improvements in signal-to-noise ratio. The C type of 1980 gave a 20 dB improvement by, in simple terms, using two B types in series. Introduced in 1990, the S type was based on an advanced professional SR (spectral recording) system. It gave an overall improvement in signal-to-noise ratio of 24 dB, and it had the advantage of being somewhat more compatible with earlier systems. However, neither of the later versions achieved wide acceptance, and the B type continues to be the mainstay of the industry.

After the mid-1970s, the manufacture of compact cassette recorders moved increasingly to Japan, to large companies like Panasonic, JVC, Hitachi, Toshiba, and Sony, as well as to a host of smaller companies. The U.S. companies that remained in the business were mostly supplied from Japanese sources. Manufacturing continued in Europe, with Philips and several German companies active.

Taken together, the tape and related improvements, and the use of noise suppression, gave the compact cassette the potential of approaching the quality of the main competing format, the LP phonograph record. By the late 1970s, the once lowly cassette was increasingly seen by consumers as having graduated to high-fidelity status.

Duplication of Prerecorded Cassettes

As the sound quality of the cassettes improved, the record companies became increasingly interested in providing the public with prerecorded cassettes, particularly as they saw the 8-track cartridge heading toward its demise as a player for the automobile. As a result, the compact cassette soon became, as well as the most popular format for consumer audio tape recording, the only tape format for which prerecorded material is widely available.

The low speed of the cassette makes it possible to duplicate cassette tape at speeds orders of magnitude greater than the playback speed, allowing for much greater efficiency and economy than is possible in the duplication of video recordings. The typical procedure was to record a master tape at, say, 7.5 in./s using Dolby B-type prerecord noise reduction. The master was then reproduced at up to 64 times the recording speed and recorded onto a series of slave machines running cassette tapes at 64 times normal speed, or 120 in./s. Usually, successive duplications take place in the reverse time sequence, so that the master does not have to be rewound. A bias frequency of about 10 MHz is necessary during recording.

Other Cassette Applications

In addition to its use as a recorder/player in the home and automobile, the compact cassette found a place in several other environments. A few of these deserve mention.

Walkman

New battery-powered portable audio devices in the late 1970s further expanded the market for the Compact Cassette. The Sony Walkman pioneered a pocket-sized cassette player that provides high-quality sound, through headphones, in a very small package. For many who had never experienced the enjoyment of music under headphones, the Sony Walkman was a new audio experience, effectively shutting out the outside world.

Boom Boxes

The "boom box" is also a battery-operated portable machine, but it provides medium-quality sound from loudspeakers. The boom box was a more social instrument, allowing owners to share their music with others (regardless of whether they were willing listeners). Young people formed the initial market, but as higher quality versions were made available they became more widely used, providing an inexpensive portable alternative to home stereo equipment.

Answering Machines

Another application that expanded enormously in the last decade is the use of cassettes in telephone answering machines. Many answering machines are based on the use of tape cassettes—often small cassettes, or "microcassettes"—

sometimes operated in pairs, one for outgoing, and one for incoming messages. Widespread use of answering machines has changed the way in which we make use of the telephone. Valdemar Poulsen would approve!

Dictating Machines

It goes without saying that the cassette, from the very beginning, dominated the dictation equipment scene, particularly in microcassette form.

Studio Cassette Equipment

The popularity of the cassette format eventually led to the manufacture of multitrack studio recorders using that format. The 4-track (in one direction) cassette decks introduced by Tascam ("Porta Studio") and Fostex in 1980 created an unprecedented interest in home recording. The decks, priced at $1300, were easily portable and included a small mixing console. Editing was impractical, since cassette tape is thin and difficult to splice; but low cost and portability more than made up for that drawback. Although not as sophisticated as studio equipment in terms of editing, these systems are capable of making recordings of acceptable quality.

CONCLUSIONS

Since the first wave of Magnetophon-derived tape recorders emerged after the second world war, sound recording on magnetic tape has appeared in a wide diversity of guises, from professional equipment the size of a washing machine to portable devices that fit in a shirt pocket.

At the professional level, analog open-reel recorders continue to be used in the radio broadcasting area, and are still used in about half of the recording studios. Digital forms of magnetic tape recording have replaced analog forms for many of the other professional applications. The takeover is, however, by no means complete.

At the consumer level, it appears that the compact cassette format will be with us for some time as the principal means of sound recording in the hands of the consumer. The compact disc now dominates the prerecorded music scene, having largely displaced the LP record. CD players are becoming more common in automobiles, so the prerecorded cassette will decline in popularity. But, as a consumer recording device, one can expect the compact cassette recorder to have more staying power.

REFERENCES

Blakely, Larry, and George, Petersen, "Multi-track Revolution," *Mix: The Recording Industry Magazine,* **6,** 46–53 (1982).

Millard, André, *America On Record: A History of Recorded Sound,* Cambridge University Press, Cambridge, 1995.

Morton, David, *"The History of Magnetic Recording in the United States, 1888–1978,"* Ph.D. dissertation, Georgia Institute of Technology, 1995.

Ottens, L. F, "The Compact Cassette for Audio Tape Recorders," *Journal of the Audio Engineering Society,* **15**, 26–28 (1967).

Philips, Frederik, *45 Years with Philips: An Industrialist's Life,* Blandford Press, London, 1978.

van der Lely, P., and G. Missriegler, "Audio Tape Cassettes," *Philips Technical Review,* **31**, 77–92 (1970).

8 The History of Digital Audio

John R. Watkinson

Of all the magnetic recording technologies, digital audio, in its short history, probably has seen the most spectacular progress. Digital audio recording is a complex technology that depends heavily on enabling technologies such as large-scale integration (LSI), which is the art of packing a great deal of electronic processing power into a small and inexpensive piece of silicon. The performance of LSI chips is subject to relentless progress, and when the degree of complexity to support digital audio became economic in the late 1970s, suitable hardware appeared almost overnight.

A chronological history of digital audio would be meaningless to readers unfamiliar with the technology and tedious to those who are. To place the history in perspective, this chapter begins with an outline of the principles of digital audio, followed by a brief chronological review of significant hardware developments. When the principles are understood, the advantages become clear, as does the significance of the various milestones.

PRINCIPLES OF DIGITAL AUDIO

In all audio recording the goal is to preserve one or more electrical waveforms, which are analogs of the motion of a microphone diaphragm. In an analog tape recording, the distance along the tape is an analog of time and the strength of the magnetic flux left on the tape is an analog of the signal voltage.

In digital audio two analogs of time and voltage are handled in a completely different way, but they remain analogs and, despite the complexity, digital audio is just an alternative way of preserving a waveform. The two axes of time and voltage are independent and can be handled in either order. The handling of the time axis determines the bandwidth or frequency response and the handling of the voltage axis determines the dynamic range, just as with analog.

Figure 8–1 shows that the time axis is handled by taking measurements at exactly uniform time increments. These measurements are called samples. The voltage axis is handled by expressing the voltage of the sample as a whole number, much like the readout of a digital voltmeter. This process is called quantizing. Figure 8–1 shows that it is possible to sample first and quantize each sample, but it is equally valid to quantize continuously and then sample the quantized values.

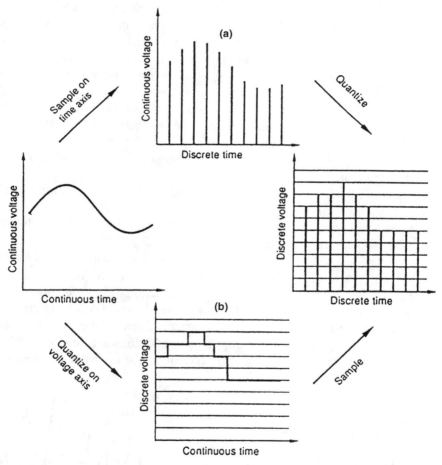

Figure 8–1   Since sampling and quantizing are orthogonal, the order in which they are performed is unimportant. (**a**) Sampling is performed first, then the sample is quantized; this is common in audio converters. (**b**) The analog input is quantized into an asynchronous binary code. Sampling takes place when this code is latched on sampling clock edges.

For recording purposes it is convenient to express the whole numbers from the quantizer in binary form. When an analog signal is converted to a binary form in this way, it is called a pulse-code-modulated (PCM) signal. PCM was invented by Reeves in 1939, when the theoretical history of digital audio effectively began.

Clearly the rate at which samples are taken affects the frequency response. Sampling is an amplitude modulation process where the sampling rate clock, or pulse train, is modulated by the audio waveform much as the carrier wave in an AM radio is modulated. Unlike an AM carrier, however, a sampling clock has harmonics, and the modulation produces sidebands around the fundamental and harmonics alike.

Figure 8–2**a** shows a correct use of sampling, whereas Figure 8–2**b** shows that when the sampling rate is too low, the waveform is incorrectly represented at a new, lower frequency. This phenomenon, known as aliasing, is due to the lower sideband of the carrier entering the baseband (Fig. 8–3). Aliasing will be familiar to anyone who has watched stagecoach wheels apparently rotate backward in cowboy movies.

Figure 8–3 also shows that an antialiasing filter can be used to limit the extent of the audio spectrum and that aliasing cannot occur if the sampling rate is a little more than twice the audio bandwidth. For an audio bandwidth of 20 kHz, which is considered adequate for high-quality audio, a sampling rate around 45 kHz might be expected.

An identical filter in a later position can eliminate the entire sideband structure, leaving only the continuous baseband signal. This reconstruction filter will be part of a digital-to-analog converter (DAC), which will be needed to reproduce a digital audio recording on conventional loudspeakers.

The quantizing process is an approximation because it forces a continuously varying parameter to be represented by the nearest step for which a whole number exists. This distorts the audio waveform at a point after the antialiasing filter and the harmonics produced can alias to produce anharmonics, an effect that can be especially distressing to the human listener because it does not occur in nature. The distortion can be eliminated by the addition of a low-level "dither" signal prior to quantizing. This signal is typically a broad-band, random signal which

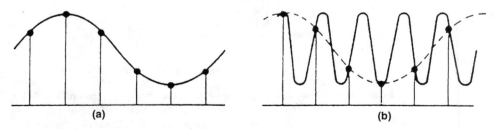

(a)   (b)

Figure 8–2   Sampling rates: (**a**) The rate is adequate to reconstruct the original
signal. (**b**) The rate is inadequate: aliasing occurs, and reconstruction produces an erroneous waveform (dashed).

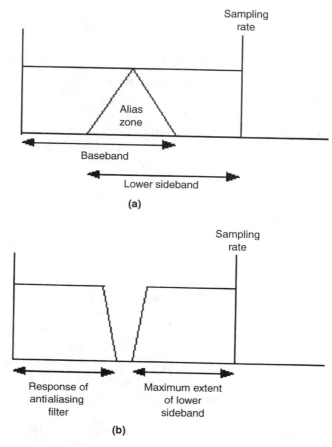

Figure 8–3 (a) Aliasing occurs when the lower sideband overlaps the baseband. This is prevented by the antialiasing filter (b), which restricts the base bandwidth to less than half the sampling rate.

decorrelates the quantizing error from the audio waveform, replacing the anharmonic distortion with a broad-band noise that is subjectively much more acceptable. Dither was first recognized in connection with video quantizing in the 1950s, but the definitive treatment of dither and audio quantizing is generally considered to be that due to John Vanderkooy and Stanley Lipshitz, published in 1984.

The number of possible values a binary code of a given wordlength can have is obtained by raising 2 to the power of the word length. Since the noise due to dither is roughly the size of a quantizing step, it follows that the more steps are available, the further the signal can rise above the dither noise. In a 16-bit system, there are 65,536 possible steps and so this is the factor by which the largest signal can exceed the noise. In audio, such ratios are expressed in decibels, in this case 96 dB. A useful rule of thumb is that 6 dB of signal-to-noise ratio is available for every bit in the word.

ADVANTAGES OF DIGITAL AUDIO

Following the conversion from analog to PCM audio, the information to be recorded is binary data and as such does not differ from any other kind of data. Essentially a digital audio recorder is a data recorder having interfaces that make it suitable for audio use. Like all data recorders, digital audio recorders can use error correction, so that if any code values are corrupted by deficiencies in the medium or by ambient noise, the original values can be restored. Once this has been done, the numbers reproduced do not differ from those that left the analog-to-digital converter (ADC).

The time axis may be corrupted by speed variations in the medium, but in the digital domain it is easy, almost trivial, to store the reproduced data temporarily in RAM (random access memory) so that the samples can be read out of the RAM with a stable clock. With this process, known as timebase correction, the original time axis of the samples can be recreated.

If the time axis and the sample values on reproduction are identical to those that existed on recording, the recording process itself has caused no loss of quality whatsoever. In principle, therefore, one of the main advantages of digital audio recording is: The quality is independent of the medium and depends only on the quality of the conversion processes to and from the analog domain. As a rule, the analog-to-digital conversion sets a limit on the signal quality, which a good digital-to-analog conversion should not make significantly worse. Of course, these conclusions assume that the medium does not cause uncorrectable errors.

When quality is independent of the medium, all the problems that beset analog recorders disappear. Wow, flutter, tape hiss, cross talk, modulation noise, and nonlinearity are all zero. Phase errors between tracks, which smear stereo images, are eliminated by reading the timebase-correcting memories of all tracks with the same clock.

In analog recording, high quality is obtained by compensating for every loss mechanism. Each compensation mechanism needs careful adjustment, and this requires a significant effort. In contrast, the digital audio recorder needs no periodic maintenance at all save for tape recorders, which require an occasional cleaning of the heads and tape path. Consequently the cost of ownership of digital audio recorders is reduced, and this feature has contributed to their popularity for professional use.

Since a digital audio recorder is essentially a data recorder, it follows that subject to economic and data rate considerations, almost any type of data recorder can be converted into an audio recorder. Similarly, digital audio recorders can be adapted to record data of other kinds. This is evident in the use of computer disk drives for audio editing, and the use of the audio DAT (digital audio tape) cassette for computer backup.

When a digital audio recording is transferred from one storage medium to another in the form of a data file, the data on the second medium are identical to those on the first and there is then no generation loss. Digital copies of this kind are often referred to as clones to underline their identical status. In contrast, analog audio recordings always suffer generation loss, with each succeeding copy having lower quality than the generation before.

Professional analog recorders must be of very high performance if they are to provide acceptable quality after successive copies have been made during the production process. This means that they are grossly overspecified for a single generation of recording. The professional analog recorder uses wide tape tracks and drives the tape at high speed, whereas the consumer recorder uses less tape in the interests of economy because the consumer is generally not interested in multigeneration performance.

In digital audio recording, the differentiation between consumer and professional is blurred because there is no generation loss. Put another way, a binary digit uses its entire information capacity in being a 1 or a 0; it cannot know whether it is a consumer bit or a professional bit.

DIGITAL AUDIO RECORDING PRINCIPLES

The analog input is converted to pulse code modulation and the data are buffered in RAM. It is universal for pages of the RAM to be read out with a faster clock than the sampling rate so that the data are assembled into blocks with spaces between. These spaces can be used to allow editing, for synchronizing patterns, and to ensure that additional check bits can be appended for error correction.

Blocks of data cannot be recorded directly on magnetic media. The data have to be modulated or channel-coded into a waveform suitable for recording. The channel-coded signal contains an embedded clock, which is used to determine the boundaries between identical bits in the serial recording. On replay this embedded clock is recovered in a phase-locked loop and used to synchronize the decoding or data separation process.

Following data separation, the data are timebase-corrected to remove any speed instabilities due to the medium, and also to remove the effect of time compression. While the data are in the RAM, the error correction process can also determine whether any data are in error and make corrections. Following error correction and timebase correction, the data can be applied to a digital-to-analog converter to recover the original analog waveform.

Error correction is necessary in digital audio because the effect of bits in error is to produce samples that can differ considerably in value from those around them. This results in annoying crackles in the reproduced sound. Errors are classified as two types: random errors, where widely spaced single bits are incorrect, and burst errors, where many bits are in error close together. A random error might be due to electrical interference, whereas a burst error might result from a scratch on the medium.

Burst errors are difficult to correct, but they tend not to occur very often and can be reduced in size by means of interleaving: samples are not recorded in their correct sequence, but are reordered prior to recording. On reproduction, the re-ordering process is reversed, with the result that the samples return to their correct sequence, but any burst errors are reordered into a larger number of small errors, which are then more readily corrected.

Error correction is an important yet arcane technology that is normally described entirely in mathematical terms. While such a treatment is inappropriate here, it is possible to get an insight into the basics. When binary numbers are used, there are only two possibilities for each symbol, 0 or 1. If a symbol is known to be in error, the correction process is the trivial one of setting the symbol to the opposite state. Consequently, in digital systems error correction becomes a problem of identifying the bits that are in error.

Figure 8–4**a** shows that the most basic way of detecting an error is to use the principle of parity. The number of 1s in the data to be protected is counted to determine whether it is odd or even. An additional parity bit is provided whose value is chosen so that whatever the data pattern, the whole word will always have even parity. Should the state of a single bit change as the result of an error, the parity will become odd, and this state can be detected by a simple count. A simple parity count detects the error but does not locate it.

Figure 8–4**b** shows that if data bits are assembled in a rectangular block, parity can be formed on rows and columns. If an error occurs, it can be located as shown at the intersection of two words that display odd parity.

In 1950 Richard Hamming devised the way of locating errors shown in Figure 8–4**c**. Here several parity counts are performed through the data to be protected, but each operates on a selected subset of the data. The reason is that should a bit fail, it will cause an odd parity condition in the counts that include that bit, but the counts that do not include that bit will appear normal. By carefully choosing the bits that appear in each count, the failure of a given bit can be made to produce a pattern of parity failures that is unique and from which the failing bit can be deduced.

The Hamming codes function graphically, but it is possible to describe how they work mathematically. The creation of the check bits from the data can be described by equations, as can the process of checking. The equivalent of locating the error at the intersection of two codes shown in Figure 8–4**b** is the solution to a pair of simultaneous equations.

The first practical burst error correcting code that worked in this way was developed by Philip Fire in 1959. Subsequently the principle was extended by Irving Reed and Gustave Solomon who, in 1960, described codes that work on multibit symbols. These codes cross the data with an arbitrary number of equations, which are paired. Solving a pair of equations produces two values:

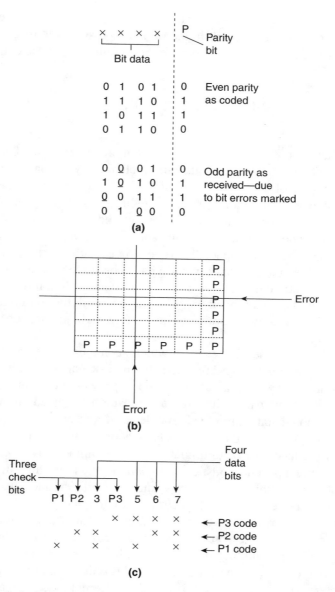

Figure 8–4 Error correction. (**a**) A simple parity check is used. (**b**) A bit in error is at the intersection of two error codes. (**c**) Use of the Hamming code makes each failing bit cause a different parity error pattern; for example, if bit 5 is in error, P1 and P3 will fail, but not P2.

one, the locator, identifies the position of the failing symbol; the other, the corrector, identifies the bits in that symbol that need to be inverted.

The Reed–Solomon codes reached the theoretical limits of correction efficiency and consequently cannot be improved on. The calculations necessary are complex and became practicable in a mass market only when the economics of LSI processing permitted. In fact it is the passing of this technoeconomic threshold that removed the last barrier to the introduction of the compact disc in 1982.

PRACTICAL DIGITAL RECORDERS

Digital audio recorders typically use sampling rates up to 50 kHz. When each sample contains 16 bits, the data rate, including synchronizing and error-correcting bits, will be of the order of a million bits per second per audio channel. A recording of one hour requires something like 500 megabytes (MB) of data. While today this requirement is taken for granted, when digital audio recorders were first being developed, this bit rate and capacity were considered prodigious in the computer industry, especially when it is considered that the bit rate is sustained and unbroken and error correction has to be performed on the fly.

For a time the high ground of high-density digital recording was held by the digital audio manufacturers. This lasted until the economics of digital video recording became favorable, when a further large step up in sustained bit rate occurred.

There are several ways in which such a data rate can be recorded magnetically. A computer-type magnetic hard disk can be made into a digital audio recorder. A RAM data buffer is used to provide a continuous flow of data as the drive itself interrupts the flow to reposition the heads between data blocks. A conventional stationary-head tape deck, or a rotary-head tape deck may be used. Recently, some digital recorders have adopted data compression or bit rate reduction to reduce cost or to miniaturize the equipment. Data compression takes advantage of a psychoacoustic phenomenon called masking, in which louder sounds reduce the audibility of quieter sounds at nearby frequencies. An audio data compressor divides the input spectrum into frequency bands. It then analyzes the input spectrum and assesses the frequencies at which masking occurs. At these frequencies sample word length can be reduced because the realized quantizing noise will be inaudible, and a saving of bit rate is obtained.

Unfortunately masking theory holds only for monophonic sound. In stereo a quieter sound is not masked if it is in a different place in the stereophonic image. Consequently on a reasonably good loudspeaker system the use of audio data compression becomes apparent as a lack of ambience or a premature termination of reverberation.

The DCC (digital compact cassette) format was launched in the early 1990s as a replacement for the consumer compact cassette and used 4:1 compression. It was not a success and, while the sound quality may have had a bearing, the main issue was that analog compact cassette hardware had become extremely low in cost and DCC appeared too expensive.

At about the same time the MiniDisc, developed by Sony, used a 64 mm diameter magneto-optical disc and 5:1 compression. Like the DCC it was not a success as a consumer audio recorder, being too expensive to compete with the compact cassette and of insufficient quality to compete with the CD.

DIGITAL AUDIO BASED ON A HARD DISK

A disk drive permits rapid access to any part of the recording. This advantage is useful for editing, since the tedious wait for tape reels to spool is eliminated. Hard disk drive audio storage was pioneered by Thomas Stockham at Soundstream, who applied it in the late 1970s to the computerized restoration of historic audio recordings, including gramophone recordings made by the Italian tenor Enrico Caruso, who died in 1921.

In 1981 the Denon company developed a hard disk recorder for its own recording studio in Tokyo, but the first commercially successful disk-based audio recorders were the AMS AudioFile and the DAR Soundstation, both produced in the United Kingdom. Since then, the falling cost of computers and disk drives has led to countless other implementations. In broadcasting, the disk-based recorder has virtually eliminated the use of analog cartridges for jingles and commercials, and is increasingly used for the editing of news material.

The cost per bit stored on a disk drive is significantly higher than that of tape, however, and so it makes economic sense to use hard disks only in applications where random access offers a true benefit. For field recording, archiving, and program playout, the time axis is perfectly linear, and the lower cost of tape ensures its survival for the foreseeable future.

STATIONARY-HEAD DIGITAL AUDIO RECORDING

Although digital audio storage requires a high bit rate, the signal-to-noise ratio on the medium itself need only be good enough reliably to distinguish a 1 from a 0. Consequently, in comparison with an analog tape recorder, the digital machine needs a higher tape speed but can use narrower tracks. A very narrow tape is weak, and the solution is to place many tracks in parallel on a tape of convenient width. In digital audio, it is quite easy to distribute data from one audio channel over several tracks so that the tape linear speed can be reduced. Alternatively, a multitrack machine may be built by using one track for each audio channel. Stationary-head machines tended to use higher sampling rates to give an extra margin of bandwidth for production purposes and to allow the use of variable speed.

In the 1970s, there were many experimental stationary-head digital machines, including an 8-track machine built by the BBC. Many of the traditional recorder manufacturers were developing digital recorders, but few reached the market. The 3M company sold a 32-track recorder having a sampling rate of 50 kHz and a linear tape speed of 45 in./s. These machines used electronic editing and so were supplied in pairs.

The most successful stationary-head format has been the DASH format (digital audio stationary head) pioneered by Toshi Doi at Sony in the late 1970s and sold from the early 1980s. This system was designed to suit many track configurations, but the greatest sales by far were in the 24- and 48-track models, which were widely adopted for record production and film soundtrack work.

Initially, DASH used a sampling rate of 50.4 kHz because it was 8/7 of 44.1 kHz, but this was changed to 48 kHz when the latter was standardized as a production sampling rate. At 48 kHz, both formats used 0.5-inch tape traveling at 30 in./s, with a pro rata reduction of speed at lower sampling rates such as 44.1 and 32 kHz.

The 24-track format used ferrite heads, and the 48-track format used thin-film heads. The DASH format, designed before the use of Reed–Solomon coding became economic, relied on a simple parity correction system allied to a very large interleave. The DASH format was supported by equipment from other manufacturers, notably Studer and TEAC.

Alongside the DASH format, a 32-track format known as ProDigi was developed by Mitsubishi, and was also available from Telefunken and Otari. In the ProDigi error correction system, an error in a given tape track was corrected using information from other tracks. Consequently it was not possible to build up a recording one track at a time unless the tape had been formatted first.

Many stationary-head stereo recorders were developed, but of these only the early Mitsubishi X-80 achieved much success. The problem was that a two-channel, open-reel recorder was both large and slow. While multitrack recording stores a lot of data, the dual-track format offers only limited advantage, and the disk-based recorders were able to erode the market with a faster access product. The market that remained was soon captured by the DAT format (see the next section), which offered identical sound quality at a fraction of the size and cost. Demand for stereo open-reel digital machines disappeared virtually overnight, with the result that manufacturers had to literally give away their recorders to educational establishments.

ROTARY-HEAD DIGITAL AUDIO RECORDING

The stationary-head recorders developed for digital audio were expensive, while the mass-produced rotary-head analog video recorder was emerging as a device with wide bandwidth yet remarkably low cost. To press analog video recording technology into service, a bizarre device known as a PCM adapter was devised; it contains much of the processing of a conventional digital recorder except that the channel coder is replaced with a system that produces a video signal output convincing enough to be recorded by an unmodified video cassette recorder (VCR). The gray scale of the video signal is replaced by two levels, one representing a binary 1 and the other representing a 0. When viewed on a TV monitor, the signal looks like an animated chessboard!

The first use of a video tape recorder in this way was in an experimental machine by NHK (Japan Broadcast Corporation) that employed a 2-inch quadruplex video tape deck and was demonstrated in monophonic form in 1967 and in stereo form in 1969. This experimental machine led to the development of practical machines for vinyl disc mastering, the first of which was produced by Denon in 1972. Again using a 2-inch quadruplex transport, this machine weighed 400 kg. It is a sobering thought to the user of today's DAT machine that a one-hour quadruplex tape weighed 10 kg and cost $500.

To produce a video signal satisfying broadcast standards, the two-level coding could be used only in the active line area, where the visible picture would normally be. For simplicity these PCM adapters placed a whole number of samples on each line, three from each channel. In the 60 Hz, 525-line system, there are 245 active lines in each field. Consequently the sampling rate was given by:

$$3 \times 245 \times 60 = 44,100 \text{ Hz}$$

The development of the compact disc led to a requirement for recorders to make the master tapes from which the discs themselves would be cut. Consequently the sampling rate of the compact disc, and the rate used in much subsequent consumer digital audio equipment, was set by the simple requirement to master on analog video cassette recorders.

Stereo PCM adapters of this kind were produced by several manufacturers. In 1978 Sony produced the PCM-100, but the workhorse recorder when the CD was launched in 1982 was the PCM-1610, which used U-Matic industrial cassette recorders. Japan Victor Corporation sold a similar unit that used VHS recorders, and in 1979 Denon brought out the DN-035R, a four-channel unit using U-Matic cassette recorders.

Consumer PCM adapters also became available toward the end of the 1970s. This advance was aided by the standardization of a format by the Electronic Industries Association of Japan (EIA-J) in May 1978. Many manufacturers produced EIA-J format PCM adapters, but the best known is probably the Sony PCM-F1. This device was compact enough to be used with a battery-powered Betamax cassette recorder to allow genuinely portable digital recording with a consumer price tag.

The EIA-J format machines lacked the flexibility of fully professional units. For the price, however, the sound quality left analog recorders in the dust. Many were bought by professionals and served as their introduction to the new digital technology.

The use of an unmodified video cassette recorder as a recording device allowed low cost but was suboptimal technically, as well as requiring two boxes. Tony Griffiths at the UK-based Decca Record Company recognized this and, in the late 1970s, developed a rotary-head digital recorder in which only the deck mechanism of a video recorder was used. The modulation scheme was a direct digital channel code onto the tape tracks. These units were used extensively in-house for vinyl and later compact disc mastering and were the forerunners of the DAT format.

DAT is a miniature rotary-head format that was announced in 1983 and immediately set a record for magnetic tape recording density. Although originally designed as a consumer digital audio recorder, the DAT format was never launched as such because of pressure from record companies, which feared uncontrolled home copying of copyrighted material. DAT was, however, immediately adopted by the professional audio industry. The DAT cassette uses flangeless spools where the edge of the tape pack is guided by lubricated liner sheets, like the analog Philips compact cassette. This system is adequate for a consumer product but not ideal for professional use.

In addition to these widely available formats, a number of other recorders were developed. For example, the Swiss Kudelski Company recently offered a four-channel rotary-head, open-reel recorder known as the Nagra-D, intended for film sound recording. Furthermore, a number of semiprofessional digital recorders are now available that record eight independent audio channels using S-VHS or 8 mm video cassettes. Since these can be synchronized to offer larger numbers of channels, the stationary-head multitrack recorder is effectively an endangered species.

CONCLUSIONS

With the first practical digital recorder demonstrated in 1967, the history of digital audio recorders is but three decades old. Even so, today we have affordable equipment that exceeds the performance of our ears. In contrast, the Wright brothers flew in 1903, yet by 1933 the biplane was still common and the turbojet engine unknown. This remarkable rate of progress has left many breathless, and there have been all too many cases of traditional skills being rendered useless by new technology.

REFERENCES

Anazawa, T., et al., *Audio in Digital Times*, AES Inc., New York, 1989.

Bellis, F. A., "A Multichannel Digital Sound Recorder," presented at the Video and Data Recording Conference, Birmingham, England: *IERE Conference Proceedings*, No. 35, 123–126 (1976).

Doi, T. T., K. Odaka, G. Fukuda, and S. Furukawa, "Cross Interleave Code for Error Correction of Digital Audio Systems," *Journal of the Audio Engineering Society*, **27**, 1028 (1979).

Doi, T. T., Y. Tsuchiya, M. Tanaka, and N. Watanabe, "A Format of Stationary-Head Digital Audio Recorder Covering Wide Range of Applications," presented at 67th Audio Engineering Society Convention, New York, 1980, preprint 1677, H6.

Fire, P., "A Class of Multiple-Error Correcting Codes for Nonindependent Errors," *Sylvania Reconnaissance Systems Laboratory Report*, RSL-E-2 (1959).

Hamming, R. W., "Error-Detecting and Error-Correcting Codes," *Bell Systems Technical Journal*, **26**, 147–160 (1950).

lngbretsen, R. B., and T. G. Stockham, "Random Access Editing of Digital Audio," *Journal of the Audio Engineering Society*, **32**, 114–122 (1982).

Ishida, Y., S. Nishi, S. Kunii, T. Satoh, and K. Uetake, "A PCM Digital Audio Processor for Home Use VTRs," presented at 64th Audio Engineering Society Convention, New York, 1979, preprint 1528.

Lokhoff, G. C. P., "DCC: Digital Compact Cassette," *IEEE Transactions on Consumer Electronics*, **CE-37**, 702–706 (1991).

Nakajima, H., and K. Odaka, "A Rotary-Head High-Density Digital Audio Tape Recorder," *IEEE Transactions on Consumer Electronics*, **CE-29**, 430–437 (1983).

Reed, I. S., and G. Solomon, "Polynomial Codes over Certain Finite Fields," *SIAM Journal*, **8**, 300–304 (1960).

Reeves, A. H., U.S. Patent 2,272,070 (1942).

Tanaka, K., et al., "On a PCM Multichannel Tape Recorder Using a Powerful Code Format," *Journal of the Audio Engineering Society*, **28**, 554–556 (1980).

Vanderkooy, J., and S. P. Lipshitz, "Resolution Below the Least Significant Bit in Digital Systems with Dither," *Journal of the Audio Engineering Society*, **32**, 106–113 (1984).

Watkinson, J. R., *The Art of Digital Audio*, 2nd ed., Focal Press, Oxford, 1994.

Watkinson, J. R., *Compression in Audio and Video*, Focal Press Oxford, 1995.

Watkinson, J. R., *The Art of Sound Reproduction*, Focal Press, Oxford, 1997.

9 The Challenge of Recording Video

Frederick M. Remley

To achieve satisfactory performance, systems designed to record and reproduce video signals must have special characteristics. This chapter identifies some of the important requirements for video recording systems and provides background information for the other chapters on video recording, which describe specific approaches to recorder design.

Television is the primary source of entertainment, cultural information, current events programs, and news in millions of homes the world over. In most countries the introduction of television broadcasting occurred not long after the end of the second world war, in 1945. By 1950 most major cities were served by at least one television station, and television services expanded rapidly. Today almost all areas of the world have some form of television service.

The successful production of television programs depends very much on the use of recording systems. Motion picture film was the first medium used for recording television programs. However, from its inception in the mid 1950s, magnetic video tape recording has supplanted the use of film recording in most television applications because this newer system offers several advantages. For example, the magnetic video recording system enables producers of television programs to use video technology from the beginning to the end of program production. This makes good sense because video production is an efficient and speedy process. No time-consuming chemical processing steps are involved, as would be required for film systems, and editing can be accomplished with ease. In fact, most films shown on television are copied to magnetic video tape before transmission. This permits more efficient operation at the television transmission facility, because handling magnetic tape is a simpler process than handling motion picture film. Despite these advantages, some elaborate and expensive television productions continue to be produced on film by means of traditional motion picture techniques, although usually editing is done only after the film images have been transferred to video tape.

Magnetic recording is the preferred means for storing video images in many other applications as well. Magnetic video recorders are found in schoolrooms, business conference rooms, and family rooms, as parts of security systems in stores and banks, and even in outer space, as components of satellites and research rockets. Many of the uses of television that we enjoy today would be inconvenient, and others would be impossible, without magnetic video recording systems. The manufacture and sale of magnetic video cassette recorders (VCRs) for home use, and the associated rental of recorded cassettes, have developed into large business enterprises.

THE NATURE OF VIDEO SIGNALS

Television signals consist of two main parts—a visual part, called the video signal, and an aural part, called the audio signal. There also is a third component, concealed in the video signal beyond the edges of the TV picture and invisible to the viewer. It contains the blanking and synchronizing information needed to control the overall television system. Figure 9–1 illustrates the kind of analog voltage signal used in traditional analog television systems.

In the real world, the senses of seeing and hearing are continuous processes involving the brain and the organs for visual sensations (eyes) and for aural sensations (ears). In a television signal there are also visual and aural components, but they are electronically produced representations, or analogs, of the real world. These analogs are obtained by use of specialized sensing systems as substitutes for eyes and ears—the video camera and the microphone.

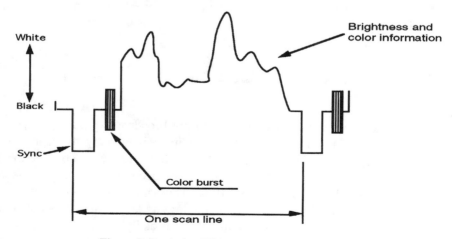

Figure 9–1 A simplified video voltage waveform.

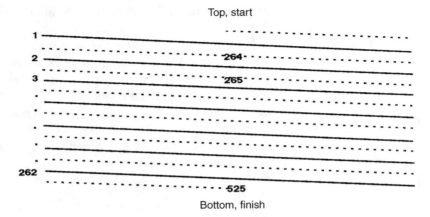

Figure 9–2   Interlaced scanning for a 525-line system.

In the jargon of present-day technology, the video signal is "sampled," not continuous. This means that a video camera, equipped with a lens and a light-sensing system, analyzes a scene toward which it is aimed and produces an electronic video signal consisting of a series of image samples spaced evenly in time. The sampling rate of an analog video signal is called the frame rate, and it is almost always either 30 or 25 frames per second (frame/s). Thus, a video signal is the sampled analog of a desired scene, just as a motion picture film contains photographic images that are image samples analogous to the scene in front of the motion picture camera. Figure 9–2 illustrates the system of interlaced scanning that is used in conventional television systems for sampling a scene to be televised.

A microphone system provides an audio signal that is a continuous analog of the sounds in the scene. For purposes of this discussion we assume that the audio signal is recorded by one of the methods described earlier in this book, and so in this chapter we investigate only the problems associated with recording video signals.

ADDING COLOR TO TELEVISION

A standardized color television system for broadcasting was approved by the U.S. Federal Communications Commission in 1954 and by the Canadian broadcast regulatory authorities immediately thereafter. However, color television broadcast services developed rather slowly, not becoming broadly available in North America until the period 1960–1964. The same system was put into service in Japan by 1960. This pioneering technology is known as the NTSC color system, deriving its name from the initials of the broadly based technical committee that prepared the fundamental specifications, the National Television Systems Committee. The complete NTSC color television signal contains, simply stated, scene brightness information nearly identical to that in an original 525-line, 30 frame/s monochrome television signal broadcast in North America since 1940, with color information added to it. The broadband signal originally called the monochrome video signal is renamed the luminance signal in the NTSC color standard. The color information is carried by an additional signal called the color subcarrier, which occupies part of the bandwidth assigned to the luminance signal. The 3.58 MHz frequency specified for this color subcarrier lies just below the upper limit of the 4.2 MHz video passband assigned to the luminance signal. The subcarrier has certain special characteristics that were carefully selected to reduce its visibility in the final television picture. The color subcarrier is also modulated in a specific way, called quadrature amplitude modulation (QAM), allowing two independent color information streams to be transmitted simultaneously with the luminance information. The two color information streams are called color difference signals, and they carry hue and saturation values for each picture area in the television scene. The NTSC luminance signal and the two color difference signals are formed in a signal processor that uses linear matrices to combine electronically specific ratios of the red, green, and blue video signal components that originated in a color television camera. In the color television receiver the luminance and the color subcarrier signals are recovered by filtering and demodulation; then a complementary matrixing process is used to recover the color signals found in the original scene.

Color television was introduced in Europe toward the end of the 1960s. Two variants of the NTSC system were approved in 1966 by an agency of the International Telecommunications Union known as CCIR (Comité Consultatif International des Radio-Communications). These systems are called PAL, for phase-alternating lines, and SECAM, for the French version of "sequential color with memory." Both transmit luminance information using specifications derived from the 625-line, 25 frame/s, monochrome broadcast systems that were already used in Europe. As in NTSC color, each of these systems uses a color subcarrier. However, the 4.43 MHz PAL subcarrier uses QAM modified by line-to-line phase reversals, while SECAM uses two adjacent subcarriers, each FM-modulated. PAL and SECAM design parameters were chosen to be somewhat tolerant of phase errors introduced by video recorders and microwave transmission systems, as well as to avoid some payments for patents used in the NTSC system. PAL and SECAM systems have provided excellent service in Europe and elsewhere.

RECORDING VIDEO SIGNALS ON FILM

Prior to 1957, during a period starting with the earliest television broadcasts in the mid-1930s, photographic motion picture film was used both as a source of program material and as the only available television recording medium. Photography, a chemistry-based analog recording system, makes use of the photosensitivity of silver halide particles that have been coated in a thin layer on a plastic film. When these particles are exposed to light rays that have been imaged on them through a lens, the particles undergo a chemical change and can subsequently be processed by chemical solutions to form visible images of metallic silver. These images are analogs of the visible elements contained in the scene. The photographic process exhibits nonlinear characteristics in the transfer from light quanta to the final silver image, and this inherent characteristic requires special consideration when film is used as a video recording medium.

For film recording of television signals, a specially modified motion picture camera, loaded with fine-grain motion picture film, photographs video signals displayed on a high-brightness cathode ray tube called a kinescope tube. Typically, both the camera and the display have been modified to minimize picture defects resulting from the almost universal practice of photographing and displaying film images at a nominal frame rate of 24 frame/s, while television systems use nominal frame rates of 25 or 30. Aside from accommodating this difference in frame rates, the photographic part of the recording process is essentially the same for television film recordings and for films intended for projection in a theater. The sound portion of the television program may be recorded on film or on magnetic material, just as in the case of theatrical motion picture use.

A processed photographic television recording is usually called a "kinescope recording" or, more familiarly, a "kine." Because of the many variables in the combined electronic/photographic process, the quality of such recordings often leaves much to be desired. Defects often encountered in photographic video recording include relatively poor image resolution; a compressed brightness range often limited by kinescope display technology to a brightness ratio of about 40:1; nonlinearity of the recordings, as exemplified by lack of gradation in both the near-white and near-black portions of the reproduced pictures; and excessive image noise due to film grain and video processing artifacts. The final signal-to-noise ratio is often less than 40 dB, especially in the case of 16 mm film.

Many of the defects just named can be traced to transfer function losses. These are incurred when the video signals representing the original scene are converted to an optical image by the high-brightness cathode ray tube, photographed by the motion picture lens/camera system, and then subjected to a series of relatively complex photographic processes. Finally, the images contained in the film recording are converted back to video by means of another video camera and some associated electronic processing as the final steps in the recovery of the stored representation of the original scene. All too often the result is a noisy picture exhibiting poor resolution and poor gray-scale quality.

Given these difficulties with film recording, combined with the imminent introduction of generally available color television services, it is not surprising that, by the early 1950s, television broadcasters and production centers were very eager to find a replacement for photographic television recording. The requirements for a replacement included:

- Excellent input signal/output signal fidelity, including a signal-to-noise ratio of 45 dB or greater and good linearity
- Simple, reliable interchangeability of recordings between users
- Easy editing, at least comparable in difficulty to editing film
- Relatively low costs for recording materials and equipment maintenance
- Instant replay

MAGNETIC VIDEO RECORDING: EARLY EXPERIMENTS

The medium most likely to meet the foregoing requirements seemed to be some form of video recording using magnetic tape. The successful introduction, beginning in 1948, of high-quality audio magnetic tape recording systems lent substance to a hope for magnetic video recorders of similar high performance and manageable cost.

Various experimental video recorder schemes using fixed heads were demonstrated in the early 1950s, but no design reached a level of performance that justified manufacture. A fixed-head approach to video recording, while theoretically possible, is difficult to implement. We can understand some of the difficulties if we consider that high-quality analog audio recording systems are required to perform over a frequency range of about 20 Hz to 20 kHz, or a span of about 10 octaves. Conventional analog video signals typically contain signal components that occupy a frequency range from about 30 Hz to as much as 5 MHz, a span of more than 17 octaves. Using fixed heads to record signals with such broad bandwidths requires high head-to-tape speeds, in the vicinity of 400 to 600 cm/s. Such high speeds result in the consumption of long lengths of tape.

Some early designs proposed recording a number of longitudinal tracks one above another along the length of a wide tape as a means of reducing total tape length. A fixed-head design of this type would require rapid shuttling of the tape from end to end. Such a machine might begin a recording on the top track, for example, then quickly reverse the tape motion as the end of the tape was reached, move the recording head down, begin recording on the next lower track with the tape going in the opposite direction, and so on. Such a design presents very difficult mechanical problems.

Other stationary-head approaches proposed the separation of the input video signal into several frequency bands, followed by shifting the frequencies within and between the bands downward, and finally recording each band separately on a multiple-track tape using wideband analog data recording technology. Tape playback would reverse this process by recombining the reproduced frequency bands after suitable frequency shifting. However, the high-speed transport elements required to implement such systems had significant operational disadvantages. Typically, for example, they were noisy. Moreover, the rapidly moving parts threatened operator safety, and the recording heads as well as the tape itself wore out quickly. Fixed-head designs were not placed into production for video use because of these and other problems, and video recorder research moved toward the use of rotating recording heads instead of rapidly moving tape.

Chapter 10 provides more information on fixed-head recording system designs.

THE FIRST BROADCAST-QUALITY MAGNETIC VIDEO RECORDER

As we have seen, various experimental video recorder schemes were demonstrated in the early 1950s, but no design reached a level of performance that justified manufacture. The first successful video recording system offered for sale was a large, complex machine developed by the Ampex Corporation in California. It was demonstrated at the 1956 conference of the National Association of Broadcasters and delivered to a few customers the following year, identified as the VRX-1000. This machine, based on an innovative recording configuration now called the quadruplex format, used 2-inch-wide tape moving linearly at 15 in./s past, and in contact with a thin wheel holding four recording heads evenly spaced around its circumference. The wheel rotated in a plane at right angles to the tape motion. A concave vacuum guide system constrained the moving tape to the shape of a 90-degree segment of the wheel and firmly pressed the cupped tape (sometimes called a tape "canoe") into contact with the four recording heads that were mounted on the periphery of the wheel. This unique mechanical configuration produced a series of short video tracks laid down across the width of the tape in the form of magnetized stripes about 1.7 inches long and 0.015 inch wide. The wheel rotated at 240 rps, and the effective writing speed was about 1500 in./s. At this writing speed it was possible to record wavelengths sufficiently short to capture and later to reproduce a specially modulated carrier signal that could accurately convey the content of a monochrome video signal. This arrangement also took advantage of the mechanical inertia of the rapidly spinning wheel to improve the timebase stability of the recorded video signal. This format was continually improved as video recording technology developed, and before long quadruplex recorders were fully capable of excellent color television recordings, as we will see.

The quadruplex recording system, and subsequent approaches to professional magnetic video recording using other rotating-head arrangements, are discussed in detail in subsequent chapters.

VIDEO RECORDERS FOR HOME USE

The design of stationary-head video recorders for home use was unsuccessful, for the reasons given earlier. By the mid-1960s all research efforts on home recording systems were directed to perfecting low-cost, rotating-head designs with the head assemblies configured in various ways. However, despite a variety of approaches in Europe as well as Japan and the United States, the mass market for home video recording could not truly develop until the introduction of competing formats using rotating heads to produce a "helical scan" track arrangement. These competing systems, which first appeared in Japan in 1975 and 1976, made use of a recording technique called "color-under" and used 0.5-inch-wide video tape mounted in cassettes. They also incorporated the latest developments in head design and in tape coating technology. In time, a single format (VHS) came to dominate the marketplace, and video cassette recorders became common household appliances the world over. Stores renting video cassettes became familiar sights in many cities, and Hollywood motion pictures eventually derived a large portion of their receipts from the sale of duplication rights to companies equipped to make thousands, often millions, of video cassette copies. As equipment designs reached progressively higher levels of performance, consumer format machines displaced more costly equipment in education and training uses. The consumer system eventually became a nearly universal choice for use in nonbroadcast video recording.

The evolution of video recording systems for home use is detailed in Chapter 13.

OPERATIONAL FACTORS AFFECTING VIDEO RECORDER DESIGN

Professional magnetic video recording systems are designed to meet the stringent requirements of television program producers and distributors. The equipment requirements of consumers of television programs—viewers in homes, offices, factories, and classrooms—are equally important, but have different emphases. And, of course, there are many more consumers than producers.

The producers of television programs are most frequently network television studios or independent production houses. Program distributors may be broadcast stations, cable systems, or broadcast satellite operators. Producers and distributors alike expect the highest quality for program recordings and seek the flexibility both to edit and to duplicate the completed recordings. Television viewers, at the other end of the system, expect that the recorded programs they watch will exhibit at least satisfactory picture quality and, in the case of home systems, will be inexpensive to rent or own and to operate. It is worth reemphasizing that as a result of improvements made in video recording, the great majority of all television programs have been recorded at one time or another before they reach the final viewer. Even motion pictures originally released for theatrical use are copied to video tape before distribution by television systems. This is done to take advantage of the consistent performance that characterizes magnetic video recording systems, as well as the possibilities offered by automatic editing and control systems designed for such equipment.

TELEVISION PRODUCTION REQUIREMENTS FOR RECORDING SYSTEMS

Entertainment programs usually have production requirements different from those typical of, for example, news programs. This is true because entertainment programs seldom plan for live action viewing by an audience, while news programs are often viewed in real time. Instead, entertainment programs depend heavily on flexible, high-quality video recording and editing systems to permit easy assembly of program segments over a period of time. Through this editing process the producer can adjust the timing and sequence of any number of scenes in preparing the final version of a production.

Television Production Processes

Most of the processes used for present-day television program production were not available before the invention of practical magnetic video recording systems in the late 1950s. We have seen that early systems were expensive and bulky; they were initially used only by television broadcast networks and a few production studios. Even so, video tape recorders were seen as important production tools from the very beginning. Production systems for televised entertainment programs make use of two fundamentally different kinds of recording technology:

- Photographic recording, where programs are recorded on film ("filmed"), often using production and editing techniques very much like those used for theatrical motion pictures.

- Magnetic recording, where programs are recorded ("taped") and edited using magnetic recording equipment to store the video and audio outputs of a television studio.

As noted earlier, most television programs are transmitted to viewers from magnetic recordings regardless of which of these methods was used to make the original recording. In addition, the programs that make use of photography for the original scenes are typically edited electronically after transfer of the film images to magnetic tape through use of a television camera system called a telecine machine. Preference for electronically editing most television programs, regardless of the source technology, results from the complexity and inevitable delays inherent in the photographic processes and the relatively fragile nature of the photographic image and its plastic support. In contrast, magnetic recordings, either copies of filmed materials or recordings from television cameras, can be viewed instantly, usually are protected from harm by a plastic cassette, can be duplicated with ease, and can be edited with greater facility than is true for film. Even in the present day production of motion pictures intended for theatrical release, the film's negatives are given final editing only after the director's decisions have been incorporated by use of an electronic magnetic tape editing system.

Program Exchange and Recording Format Standards

Professional television program production and distribution organizations frequently exchange magnetic television recordings. Home and business users expect successful recording exchanges as well. Thus there is a requirement to achieve and maintain interchangeability of recordings between machines made by different manufacturers and located in different places. This, in turn, results in a need for format standards that define the electrical, mechanical, and magnetic characteristics of the recordings that are to be interchanged.

A variety of factors must be addressed to assure the success of magnetic recording interchanges. The magnetic tape, including both the plastic backing and the magnetic coating, together with the reel or cassette that carries the tape, must be mechanically stable over time and must accommodate temperature, humidity, and atmospheric pressure variations in predictable ways. These requirements must be met not only in the short term, soon after recording, but also after long periods of storage.

Careful recording equipment designs, incorporating robustness and stability, are mandatory to guarantee recording interchangeability. Several kinds of mechanism, each with its own performance requirements, are needed for the operation of video recording systems. Examples include the mechanisms for loading the tape cassette into the machine at hand and for moving the tape accurately to the proper position relative to the magnetic heads. Included also are the drive system responsible for moving the magnetic medium past the head assemblies, and the system for driving and controlling the speed and position of rotary heads. There is an implied overall requirement for each of these mechanical systems to accommodate the normal wear that occurs during use, as well as to provide satisfactory performance during operation under conditions of varying temperature and humidity.

In the early days of magnetic video recording some of these requirements for interchangeability had not been fully identified. Even when known, several factors were difficult to bring under control. To allow adjustment for some mechanical variables, and for ease of replacement, one of the basic design features of all quadruplex recording machines was provision for a removable video head subassembly that contained the rotary-head wheel drive and concave tape guide, the main elements determining video track geometry. During the pioneering days of the format it was common practice, when exchanging recordings, to supply not only a recorded tape but also the expensive video head assembly that was used during its recording. This situation posed severe operational and economic penalties on tape exchange and led to a concentrated effort to control the variables in the system that most affected interchangeability. Significant successes were achieved in a relatively short time, as the details of the process came under closer scrutiny. In some cases better materials could be chosen for components and, overall, better control and assignment of tolerances were achieved during the manufacture of the most critical components.

We can draw a clear distinction between a standardized recording format and a specific make or model of machine that makes recordings conforming to that format. A recording format specifies the magnetic, mechanical, and physical characteristics of a recording. Recording equipment for use with a specific format must be constructed to record and reproduce (play back) recordings that meet, within specified tolerances, the specifications of the format. Most television recording formats are standardized by national or international standards groups. In the United States, the Society of Motion Picture and Television Engineers (SMPTE) prepares video recording standards that are approved and published by the American National Standards Institute (ANSI). International standardization of video recording formats is accomplished within Technical Committee 100 of the International Electrotechnical Commission (IEC TC 100), and the IEC publishes a comprehensive group of format standards documents.

It can be noted that the ongoing requirement for recording interchanges among groups of users may result in the relatively slow evolution of standardized recording formats. This is especially true at the program production level, since long-term interchangeability of recordings is vital to the operation of production organizations. Thus, recording technology may advance more swiftly than standardized recording formats can change. For example, the careful design choices of the first magnetic video recorder used in broadcasting, the quadruplex recording system, were internationally standardized in the early 1960s. These quadruplex format recorders continued in general use in television production studios and broadcast stations for nearly 20 years, with only minor format refinements.

In the middle of the decade of the 1970s, however, the SMPTE Type C and Type B formats were standardized. These efficient formats provided superior performance with other benefits, including cost reductions both for equipment acquisition and for operation. They replaced the quadruplex format within a few years. The time was certainly right for a major technical step forward, but the new formats offered not only high levels of performance but also savings in purchase price and cost of operation. Because of these several factors, the historical requirement for interchange of both new and old recordings was reduced to a low level of importance. A similar transition occurred in the 1980s, when professional digital recorders were introduced to the program production and distribution industries and provided improved multigeneration performance, thus increasing the manageability of complicated editing procedures.

The continuously falling costs for digital electronic components, coupled with advances in video recorder designs, assure the availability of affordable home-use digital recorders in the near future. However, considering the examples noted above for professional equipment, we can expect a relatively long transition period before digital recorders replace the millions of analog video recorders found in homes around the world.

SUMMARY

Present-day video recording systems encompass a wide range of applications, from costly master-quality recorders found in the studios of production centers, to lightweight portable systems used by news crews, to the inexpensive video cassette recorders found in millions of homes and classrooms. The most recent improvements in video recording technology have applied digital techniques to the special requirements of television production and distribution. The descendants of the early analog video recording systems have spread throughout the world, but will be replaced gradually by digital video recording machines that exhibit better performance.

The details of most of the video recording formats mentioned here will be found in the following chapters. These chapters also present interesting descriptions of the many unsuccessful formats that were proposed in the course of four decades and more of video recording. Today, the evolution of recording formats continues at a brisk pace. New challenges lie ahead in meeting the recording needs of new, advanced television systems, including the high-definition systems that are expected to dominate viewing in the twenty-first century.

REFERENCES

Benson, K. B., Ed., *Television Engineering Handbook*, McGraw-Hill, New York, 1986.

Fink, D. G., Ed., *Electronics Engineers Handbook*, McGraw-Hill, New York, 1975.

Jackson, K. G., and G. B. Townsend, *TV & Video Engineer's Reference Book*, Butterworth-Heinemann, Oxford, 1991.

Mee, C. D., and E. D. Daniel, Eds., *Magnetic Storage Handbook,* 2nd ed., McGraw-Hill, New York, and IEEE Press, Piscataway, NJ, 1996.

Poynton, C. A., *A Technical Introduction to Digital Video*, Wiley, New York, 1996.

Van Valkenburg, M. E., Ed., *Reference Data for Engineers: Radio, Electronics, Computer, and Communications*, SAMS Publishing, Carmel, IN, 1993.

10 Early Fixed-Head Video Recorders

Finn Jorgensen

THE MAIN PLAYERS

Videotape has become a key product in providing storage and replay of all kinds of visual material. There are today an estimated 400 million video tape recorders. Not many of those who use or enjoy these recorders are aware that the first of them was born in Tin Pan Alley. It started in 1948, in a small office/laboratory near the corner of Melrose and Gardner in Hollywood, California. The efforts were sponsored by the crooner Bing Crosby. In June 1951 he rewarded the efforts of the technical staff by moving them across town to 9030 Sunset Boulevard, into Bing Crosby Enterprises, a laboratory dedicated to the advancement of magnetic recording.

The first player in the video tape recorder development race was John T. (Jack) Mullin, an engineering graduate of Santa Clara University. He served in the Royal Air Force in England during the second world war, and was among the first to listen to the high-fidelity music transmitted from Germany during the war. The device that made this possible—a studio version (Model R22A) of the German tape recorder called the Magnetophon—used a recording tape made by BASF, consisting of a plastic ribbon impregnated with iron oxide powder. The recording head field was augmented with a high-frequency, ac bias signal, which assured a recording of extraordinarily low distortion and low noise.

Two of these machines were war memorabilia and were shipped in 18 packages to Jack Mullin's home in San Francisco. He was then with the U.S. Army Signal Corps, and returned to the United States carrying an additional package that contained the precious magnetic heads for the two Magnetophons. He put all the components together at his new place of work, W. A. Palmer Studios, and proceeded to demonstrate the capabilities of the Magnetophon. Two parties became interested: Alexander M. Poniatoff of Ampex Corporation, and the entertainer/crooner Bing Crosby, who had formed Bing Crosby Enterprises.

Ampex, which had made small electric motors for aircraft during the war, was experiencing declining business and needed a new product. The Magnetophon demonstrated by Mullin appeared attractive. It contained electric motors, and associated mechanical parts. But there were two new components: the magnetic heads and the magnetic tape. Ampex engineers started work and experimented in building heads from stampings of mumetal, a high-permeability nickel–iron alloy, and eventually they succeeded in building heads that outperformed the original Magnetophon units.

The popular film and radio entertainer Bing Crosby had been permitted to prerecord his radio show for the ABC network on 16-inch, 33⅓ rpm, lacquer discs, which were edited by rerecording to produce his 30-minute programs. The editing, however, resulted in high distortion and noise. Mullin demonstrated the superiority of the Magnetophon and, in the summer of 1947, took over the recording and editing of Crosby's radio show. Mullin's meager supply of German tape was soon bolstered by tapes from 3M and Audio Devices, as these companies envisioned the great potential in the future of magnetic recording.

The successful application of the Magnetophon to the Crosby radio broadcasts spurred Ampex in all haste to introduce their first professional audio recorder.

By late 1947 tapes recorded on the Magnetophon could readily be played back on the new Ampex machines, but the record circuitry in the Model 200, the first Ampex commercial system, had not been fully developed. Jack Mullin promised the ABC chief engineer that this was an easy situation to remedy. On his word, the ABC network ordered eight recorders. Ampex could not finance this first order, so Bing Crosby Enterprises (BCE) advanced the electronics firm $50,000 to build 20 professional audio recorders. This started a business relationship between BCE and Ampex in which BCE was sole distributor for Ampex. Soon the audio market place exploded, and Ampex set up additional distributors.

Further developments in the audio field were left with Ampex. Crosby was so pleased with the outcome of his investment in audio recording on tape that he decided to invest in video recording on a longitudinal recorder. The effort was managed by Frank Healey with technical team members Jack Mullin and Wayne Johnson (Fig. 10–1). Their work started in 1950, since as long ago as April 1948 Jack Mullin had studied the possibility of recording the video signal, and they had retained electronic genius Wayne Johnson to help in this area. Johnson had great experience in television work, having built the studio equipment for station KFI-TV in Los Angeles. Ampex did not authorize a video recorder project until 1951. Their efforts resulted in the rotating head recorders revealed in 1956, and described in Chapter 11.

The Radio Corporation of America (RCA) and the British Broadcasting Corporation (BBC) both entered the game of the video tape recorder in the early

Figure 10–1 The chairman of the board of Bing Crosby Enterprises (extreme right) is brought up to date on video tape recorder technical development by (left to right) John T. Mullin, chief engineer; Frank Healey, electronics division executive director, and Wayne R. Johnson, TV project engineer. [*Photo courtesy of Jack Mullin.*]

1950s. It appears that RCA was involved in the spring of 1953, while the BBC announced work on their Vision Electronic Recording Apparatus (VERA) in 1952.

A video signal has a set of characteristics that are completely different from an audio signal. One of the problems in recording and playing back video signals is the tremendous bandwidth required: 18 octaves compared with 10 for an audio magnetic recorder, with the highest frequency extending to several megahertz rather than some 20 kHz for audio. Moreover, a video recording must contain additional signals for synchronization of the display and positioning of the many lines that make up a video picture, and information regarding color (hue and intensity) must be included. The video signal-to-noise ratio must be equal to (or better than) 40 dB, and accurate timing of the sync signals must be maintained. Finally, the recording must include the audio portion of the television program.

As described in Chapter 9, there are several ways in which signals can represent video and its color content—Which one to use? The first attempts involved recording black-and-white video plus sound. This was achieved in 1951 and 1953, with demonstration equipment running at 360 in./s and a bandwidth that was slightly above 1 MHz. It was realized, however, that the inclusion of color adds a relatively large improvement to the pictures quality, in spite of a smaller color signal bandwidth.

The National Television Systems Committee (NTSC) composite video signal consists of a black-and-white luminance signal (Y) with bandwidth from zero frequency (dc) to 4.2 MHz plus a 3.58 MHz subcarrier (C), modulated by two color signals, I and Q, having 90-degree phase displacement. The C signal's amplitude represents the saturation level (intensity), while its phase contains the hue (tone). Any phase error (a few degrees only) will change the image color and, in the early days, the NTSC was nicknamed "never twice the same color." The bandwidths of the two signals that modulate the color carrier C have lower bandwidths than the luminance Y: the bandwidth is 0.5 MHz for the I signal and 1.5 MHz for the Q signal (Fig. 10–2).

These bandwidths are considerably wider than existed in the Magnetophon-derived recorders. The audio spectrum covers the region from 20 to 20,000 Hz, equivalent to 10 octaves. The video signals requires an additional 8 octaves at the high end plus a response that goes to dc. The resulting requirement is 18 octaves plus dc, clearly exceeding the range of ordinary playback heads that can be designed to cover up to 12 octaves ($500–2 \times 10^6$ Hz at 120 in./s).

There are several options for recording of this wideband video signal, including color, synchronization, and audio signals. How to divide the information up into what channels? Apply AM, FM, or PCM? Which signal combination?

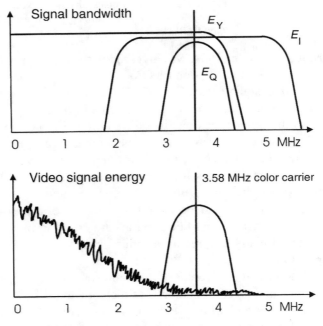

Figure 10–2   Nominal bandwidths of video signal components.

(Modern digital broadcast video has two main standards: D-1, which records digitized channels of Y, I, and Q, and D-2, which has Y and C digitized.)

As indicated in Chapter 9, the European standard PAL (phase alternating lines) is a composite signal similar to NTSC, but the color difference signals U and V reverse in phase on alternating lines, and phase errors tend to cancel. The result is a picture with far less sensitivity to phase errors. This philosophy was furthered in 1965 by a second European standard for sequential color with memory, known by the French acronym SECAM. With SECAM, the color difference signals U and V are sent separately on alternate lines, and frequency modulation of the subcarrier is used instead of amplitude and phase modulation. An application of this scheme was used in one of the Fairchild/Winston experimental video tape recorders described later.

APPROACHES TO RECORDING A BLACK-AND-WHITE VIDEO SIGNAL

In August 1951 BCE announced the development of a 12-channel recorder/reproducer, and RCA's chief David Sarnoff indicated interest in video tape recorder development. This encouraged Bing Crosby Enterprises to stage a showing of a recording on November 11. These progress announcements became regular during the next five years, often followed by demonstrations.

At first Mullin and Johnson were not ready, however, and had to resort to the original video tape recorder prototype, a modified Ampex 200, using standard 0.25-inch audiotape. The recorder had been modified by BCE in 1950. The speed was increased to a high value of 360 in./s in the hope of increasing the high-frequency response to a value sufficient to give a crisp video picture. Unfortunately, the tape surface was quite rough, producing a large effective spacing loss between the read head and the tape. There was also a drawback to the excessive speed of 360 in./s, namely, the formation of an air film between the tape and heads. Today we know that this film is formed somewhere between 60 and 200 in./s, depending on tape tension.

Initially, the lack of dc response was overcome by careful use of dc restoration circuits, borrowed from television receiver designs. Later experiments used FM modulation of the dc and the low-frequency portion of the video signal.

The premature demonstration went on, however, although the picture quality was inferior. A bit of the Hollywood atmosphere at that time is caught in the following exchange, reported by Jack Mullin to the author in 1997.

"We had 'recorded,' if it could be called that, some TV pictures of airplanes landing and taking off," Mullin recalls. "When we gave the demonstration, Frank Healey would stand by the monitor and say,

'Now watch this plane come in for a landing' or 'There goes a guy on takeoff.' It is doubtful the viewer would have known what he was seeing without this running commentary."

On the basis of this "demonstration" Frank Healey was bolder than Sarnoff and predicted that BCE would have commercial models in general use within a year. Regardless of Healey's hucksterism and the poor quality of the picture, it was the first public demonstration of television recorded on tape. Healey had achieved his publicity objective: "one-upping Sarnoff and RCA."

On November 12, 1951, an article in the *San Mateo Times* quoted Ampex head Poniatoff's announcement of his company's goal to produce a video tape recorder in 6 months. The initial budget was $14,500. The project would be led by Charles Ginsburg, hired in January 1952, later to be joined by Ray Dolby and others.

Figure 10–3   Black-and-white picture quality of playback (left) is comparable to original signal source (right), 1953. Video was recorded and reproduced with the 12 tracks system. [*Courtesy of Jack Mullin.*]

BCE came in first, on October 3, 1952, with a demonstration of a high-resolution recording of a motion picture (Fig. 10–3). Three months later both the *Wall Street Journal* and the *New York Times* reported a successful demonstration of an improved recorder, now using a tape recently developed by Minnesota Mining and Manufacturing Company (3M). The reaction to the taped Jack Benny show was varied. The press in general agreed with the statement of BCE manager Frank Healey, *Broadcasting Telecasting*, that quality was "more than 20-fold" better than the first showing in November 1951. The announcements also empha-

Figure 10–4 RCA video recorder/reproducer: reproduce and recording amplifiers are mounted in racks to the left and the right; transport for 0.5-inch wide tape on 17-inch reels. [*From Olson et al. 1954.*]

sized the obvious merits of tape recording over photographic film: no development time, and possibilities for instant replay and immediate retakes. No more rushes and one-day delay for the development of film, plus the attractive advantages of lower cost and reusability.

RCA's interest was now stirred and, on March 13, 1953, Sarnoff visited BCE for a demonstration of the recorder. He later announced that one of his three wishes for birthday gifts from his staff was a video tape recorder. Later that year RCA announced a recorder that used 0.25-inch-wide tape with one channel of video and one channel of audio (Fig. 10–4).

IMPROVED RECORDERS APPEAR IN 1954, WITH COLOR

In March 1954 both RCA and BCE presented technical papers on their video recorders in New York, at the annual convention of the Institute of Radio Engineers (the forerunner of IEEE). Jack Mullin described BCE's recorder and Harry F. Olson RCA's. Neither system employed any unusual signal processing technique; rather, both relied on brute force signal recording techniques. The limited bandwidth necessitated the use of multitrack recording. For the three organizations, the arrangements were as follows:

BCE

0.5-inch-wide tape on 14-inch reels, running at 100 in./s

10 tracks of multiplexed video, each with a bandwidth of 0.39 MHz

1 track for the video sync

1 track for FM audio

RCA

0.5-inch-wide tape on 17-inch reels, running at 360 in./s

1 track for the video luminance signal plus sync

3 tracks for the video color signals: red, green, and blue

1 track for FM audio

BBC

0.5-inch-wide tape, 20-inch spool, 200 in./s

2 tracks of video (bandsplit), plus sync: dc to 100 kHz (FM), 0.1–3 MHz (AM)

1 track for FM audio

Figure 10–5 The BBC's VERA. [*From Axon 1954.*]

The BBC's VERA is shown in Figure 10–5.

The BCE recorder used a sampling scheme to divide the incoming video signal into 10 channels. The sampling also switched the signal polarity, and this aided in handling the dc content. The recorder mechanism developed into a recorder console (Fig. 10–6, left; the five racks to the right contain electronics).

TECHNOLOGY ADVANCES, 1947–1956

Several developments during the first half of the 1950s led to improved magnetic recording performance.

Figure 10–6   1955 version of BCE's video tape recorder. Tape transport console is to the very left, followed by video playback, video recording, sound, and auxiliary electronics in five racks. [*Courtesy of Jack Mullin.*]

Better Tape Handling

Tape handling was greatly improved by employing a tight loop capstan drive in which the tape lies against each side of a single rotating capstan shaft, clamped against the capstan surface by rubber pucks (an illustration is given in Chapter 20 on instrumentation recording). This arrangement provides isolation of the tape inside the tape loop, with resulting reduction of flutter and disturbances from the reeling system. This type of capstan drive was employed by BCE and the BBC.

None of the longitudinal recorders provided acceptable pictures, scrape flutter remaining the Achilles' heel. Accurate reproduction of vertical lines in pictures was doomed because of wow and flutter. A simple geometry consideration shows that the duration of a single scan line is 63.5 μs (525 lines, 30 times per second). Of this, 52 μs is allocated to the video information, and 11.5 μs to the front and back porches plus sync signal. The duration of a single pixel display is equal to the duration of a picture line divided by the number of pixels per line, or 52 μs divided by 700. The duration of a single pixel is therefore 74 ns.

The time displacement errors (TDEs) in ordinary longitudinal tape transports are on the order of ±1000 ns peak. Timing errors of this magnitude will lead to geometric errors of ±1000/74 or ±13 pixels in a vertical line. The result is a most unpleasant picture. The current standard calls for time displacement errors of less than ±3 ns for broadcast quality, less than ±30 ns for home systems.

RCA estimated the maximum flutter to be 0.004% at a 30-cycle rate. Add intertrack error in a multichannel video recorder, and the picture quality becomes mediocre. (Intertrack error due to skew can amount to ±300 ns between adjacent tracks in an instrumentation recorder operating at a speed of 120 in./s.) Even the most clever and careful mechanical engineering has not produced a conventional tape transport mechanism that has sufficient low TDE. Only low-inertia capstan systems may succeed. RCA implemented a moving-head, rotational vibrating system that could operate at a reasonably high frequency.

The high tape speeds needed to record short wavelengths at high frequencies presented a problem during start-up. The first unit from BCE (in 1951) used 360 in./s, and so did RCA. It was an experience to watch these machines starting up. With the capstans running at full speed, the operator would press the start button to start the reel motors. By sensing the motors' rotational speed, together with the amount of tape on a reel, the operator could easily tell when the speed was within a few percent of the target 360 in./s. At that instant, the capstan pinch roller would engage full force and clamp the tape to the capstan. Everything then was supposed to go smoothly. But this was not always the case. A wrong move would result in a loud BANG plus a spectacular amount of tape spilled onto the floor!

Phase Equalization at High Frequencies

Bing Crosby Enterprises capitalized on the video experience of Wayne Johnson to employ true phase equalization of high frequencies. These frequencies correspond to short recorded signal wavelengths, which is where the major losses in magnetic recording occur. An equalization boost is therefore necessary, and ordinary circuits provide this, but with a simultaneous large phase shift. The problem is similar to one in video transmission, where a needed boost in high-frequency video signals must be accomplished without phase shift. RCA developed a delay-line equalizer to supply this boost. BCE used this so-called *aperture equalizer* in the video playback circuits and produced video playback pictures in the early 1950s that were free from phase distortion errors.

Improved Heads

Heads were developed into high-quality multitrack configurations, and core lamination material was developed to withstand the severe abrasive forces that occur at high speeds and high tape tensions. Two alloys were used: a binary alloy called Alfenol (**Al–Fe**, **N**aval **O**rdnance **L**aboratory), and Sendust, a ternary alloy (iron–aluminum–silicon), named after the university in Sendai in Japan that played a role in its development.

Gap lengths were controlled by cleaved mica sheets. Their thickness should be on the order of the wavelength of light, and an astute mechanic at the BBC developed an optical measure for sizing the thickness of the mica by judging the color of light transmitted through the sheets.

FROM THE MID-1950s TO NOW

BCE, after developing a black-and-white recorder, took the next obvious step to a color video tape recorder, demonstrating an engineering model on February 7, 1955, essentially copying what RCA had done earlier. According to Kirk (1979), Mullin stated:

> In order to record color, we abandoned the multiplex system and went to direct high-speed analog—at 360 in./s and soon thereafter 240 in./s—five tracks on 0.5-inch tape: Red, Green, Blue, sync, and FM sound. Sound was superb. Picture was actually very good. CBS was about to give us an order for machines when Ampex showed their quad system with FM video—and wiped us out!

RCA had, in the meanwhile, changed to 0.25-inch tape with two channels, video and audio. Thus Sarnoff's company was now where BCE started, and vice versa.

The end of longitudinal video tape recorder development came in April 1956, when Ampex showed the Mark IV quadruplex machine, the first broadcast video recorder with a transverse-scanning head. The picture was not perfect, there was noise, and "venetian blinds"—but these defects could be fixed. One major improvement was the absence of serious timebase errors due to the flutter inherent in the longitudinal transports.

Aiming to capture the rapidly growing market in space activities with missiles, rockets, orbiting capsules, and so on, all requiring recording equipment for telemetry, the 3M Company acquired BCE in September 1956. The venerable 11-track video tape recorder changed into a 14-channel instrumentation recorder that eventually would record 2 MHz bandwidth signals on each track. For that evolution the reader is referred to Chapter 20.

The desire to record video longitudinally did not yet disappear, since the new helical-scan home and industrial video tape recorders were quite expensive, typically costing $2000 and up when they first appeared. A longitudinal tape transport can be made more simply than a scanning-head recorder. A first home video tape recorder was developed as a "simplified video tape system for home use" at RCA in 1956. RCA built the video recorder in a color TV receiver as a "Hear-See System," which looked like a big audio tape recorder. The recorder used 0.25-inch tape on a 7-inch open reel and played back 16 minutes of color program in two passes. This video tape recorder was not pursued to a successful commercial level.

Efforts to make a longitudinal video tape recorder also persisted in Britain, where the BBC showed the VERA in 1958, for the first and last time.

Soon other companies joined the race toward a low-cost, longitudinal video tape recorder. Akai (VX-1000), Ampex (VR-303), Armour Research Foundation, General Electric, Magnecord, 3M, Shibaden (SV-1000), Webcor, and Wesgrove. And in 1962 came the Telcan unit from the Nottingham Electric Valve Company in the United Kingdom, a video recorder at the very low price of about $160. Not surprisingly, Telcan worked very poorly, if at all. It was not a completed product, but a kit from which the average television enthusiast supposedly could assemble a recorder. The unit had 0.25-inch-wide tape on an 11-inch reel and could record 32 minutes of black-and-white video in four passes. The tape speed was 120 in./s and the picture (if any) was very poor owing to low resolution and high noise. A good hoax, it seemed, but it nevertheless started another video tape recorder project at Fairchild Camera and Instrument's Winston group, where Wayne Johnson was now chief technical officer.

By then there had been some improvements to transports, heads, and tapes. A new tape, using chromium dioxide as the magnetic material, was available in small quantities from Du Pont, and much had been developed in the field of video signal processing. It was decided to give the longitudinal video tape recorder another try. Experiments showed that the new chromium dioxide could extend the response from 2 MHz to 3 MHz, which would be considered acceptable. The design goal was a recorder that used a simple tape transport, moving a 0.25-inch-wide tape at a speed near 120 in./s. A 10.5-inch reel could then hold 15 minutes of playing time. Motion reversal could take only 4 seconds—and the recording would be carried on one or two narrow tracks, which meant sharing for video, sync, and sound. The addition of color was a requirement.

In one approach the video signal was stripped of the porches and sync signal, and a new sync signal was generated as a doublet inserted in their place. The doublet occurred at a repetition rate of 15,750 Hz and, at those precise instants, the sound would be sampled and used to amplitude modulate the doublet. When demodulated upon playback (a simple low-pass filter) a high-quality, 8 kHz bandwidth audio signal resulted.

The video and color information was processed in a way that was inspired by SECAM. With SECAM, the color difference signals (R − Y) and (B − Y) are sent separately on alternate lines. During one line, (R − Y) signals are transmitted and the (B − Y) information is discarded, and on the next line (B − Y) signals are transmitted and the (R − Y) information is discarded, in a continuous sequence. In the receiver, by means of a delay line and electronic switch, a combination of past and present signals is achieved whereby the sequential information is displayed simultaneously. This results in a reduction of vertical chrominance.

SECAM inspired Johnson to a scheme in which the video recording was made up of alternating sections of (B − Y = U) and (R − Y = V), in sequence. A SECAM receiver was made part of the video tape recorder, to serve in playback. A delay line (of 63.5 µs) would provide a delayed signal path to ensure that B − Y and R − Y were detected together, providing Y, B, R and, from these components, also G. There was a reduction of vertical chrominance, but the method provided a satisfactory video picture.

The transport mechanics were improved using dual, differential capstans that did not rely on reel tension to provide tape tension over the heads (inside the capstan loop). Low-frequency scrape flutter was reduced by placing a roller that contacted the tape in the immediate vicinity of the head. The free tape span was less than a half-inch long. Further reduction of scrape flutter was achieved by letting the tape run over a Teflon post. Enough molecules would apparently transfer to the surfaces of the tape (and the head) that the Teflon-to-Teflon molecule affinity served to greatly reduce the stick–slip action.

Even so, timebase errors, caused by scrape flutter, limited the performance. A special low-TDE tape transport might have been feasible, and large consumer product manufacturers such as Sears were interested. However, no sooner were plans being drawn than the public's interest turned toward color television, and the Winston recorder was abandoned.

The camcorder was still years away, so it is not surprising that the pursuit of a longitudinal, fixed-head recorder went on. In 1965 Ampex introduced the fixed-head VR-303, using 0.25-inch-wide tape operating at 100 in./s. The bandwidth (at − 3 dB reference points) was 250 Hz to 2.5 MHz. Picture instability due to timebase errors was again apparent, as always in these fixed-head machines, and Ampex decided to withdraw the unit. In its place came a two-speed 1-inch-tape, helical-scan recorder.

In 1966 Camras showed Armour's video tape recorder, which had a tape cassette configuration containing an endless tape loop. One hundred tracks could be recorded, either in a continuous serpentine-like pattern, providing a playing time of 100 times 30 seconds (50 minutes), or in 100 separate video/audio announcements. This unit never got beyond the prototype stage.

Next Akai announced a fixed-head model VX-1100. The tape speed was only 30 in./s and the bandwidth was claimed to be 1 MHz, primarily due to the use of Marvin Camras's cross-field bias record head. Audio and video were recorded on separate parallel tracks along one half of a 0.25-inch-wide tape, and the 10.5-inch diameter reel was then turned over for another 48-minute run. The VX-1100 never appeared, but a speeded-up (45 in./s) version subsequently made it to the New York Consumer Electronics Exhibition of 1967. Thereafter nothing further was heard of it. .

The Newell tape transport was a well-promoted invention in 1966. It was a high-speed tape transport wherein the two reel packs (no flanges) are pressed against the rotating capstan and the tape moves from one reel to the other at high speed and low flutter (since there are no free tape spans to produce vibrations). Very high tape speeds became possible, and the time required to reverse tape speed was minimized by employing a high-torque, low-inertia printed circuit motor. The resulting reversal time was then merely a few milliseconds, too small an interval to be noticed by the viewer. The Newell transport was embraced by BASF in Germany, and the linear video recorder (LVR) was launched in 1979. The system employed a hundred parallel tracks across one inch of tape and claimed a bidirectional turnaround of a tenth of a second from 120 in./s, forward to reverse.

In 1979 Toshiba developed a fixed-head video cassette recorder that had lower tape consumption than a standard 2-hour VHS cassette recorder. The secret was an endless tape mechanism, which would play back 17 seconds of pictures 220 times, with a total recording time of 1 hour. The Japanese company claimed that the mechanism of its recorder was simpler than that of a rotary-head recorder, and convenient for random access.

In practice, none of these systems was able to compete with the helical-scan approach, and all were soon abandoned. Nevertheless, reports from France indicate that Thompson is developing a fixed-head video tape recorder applying new thin-film head magnetics in combination with magneto-optical technology. The reader may rest assured that somewhere, someone else is secretly tinkering with a novel longitudinal video tape recorder.

ACKNOWLEDGMENTS

The author thanks Jack Mullin for several most enlightening contributions, historical facts, and conversations on his video tape recorder experiences. I also thank Vern Bushway for help in recalling the past, in particular the 0.25-inch video tape recorder versions he helped to advance.

References

Axon, P. E., "Electronics Recording Apparatus," *Journal of the Television Society of the U.K.*, 399 (1954).

Ginsburg, C. P., "The Birth of Videotape Recording," *Ampex Monogram*, March 1981.

Healey, Frank, "Taped TV," *Broadcasting/Telecasting,* 77–78 (1953).

Johnson, W. R., and A. Ispas, "Winston Industrial Television Magnetic Tape Recorder/Reproducer," *Winston brochure,* 1964.

Kirk, D., "The History of Formats," *Television & Home Video*, 55–59, April 1979; 46–51, July 1979; 59–61, January 1980.

Lindsay, H., "Magnetic Recording, Part I," *dB Magazine*, 37–44, December 1977.

Lindsay, H., "Magnetic Recording, Part II," *dB Magazine*, 40–44, January 1978.

Mullin, J. T., "Video Magnetic Tape Recorder," *Tele-Tech,* **13,** 127–129 (1954).

Mullin, J. T., "The Birth of the Recording Industry," *Billboard*, **52**, Nov. 18, 1972.

Mullin, J., "Archive—The Crosby Video Tape Recorder," *VIDEO*, **8,** 28–31 (1982).

Olson, H. E., W. D. Houghton, A. R. Morgan, J. Zenel, M. Artz, J. G. Woodward, and J. T. Fisher, "A System for Recording and Reproducing TV Signals," *RCA Review*, **15**, 330–392 (1954).

Sadashige, K., "An Overview of Longitudinal Video Recording Technology," *J. SMPTE*, **89**, 501–504 (1980).

Stanton, J. A., and M. J. Stanton, "Video Recording: A History," *J. SMPTE*, **96**, 253–263 (1987).

Wolpin, S., "The Race to Video," *Invention & Technology by Ampex Corp.*, 52–62, Fall 1994.

11 The Ampex Quadruplex Recorders

John C. Mallinson

This chapter contains a detailed account of the early development of the 1956 Ampex quadruplex video recorder, the VR-1000, together with a less detailed review of the refinements subsequently incorporated in later models such as the VR-2000.

Quadruplex video tape recorders were the worldwide standard for professional video for about 25 years (1956–1980), with some remaining in use even today. Their initial service was simply that of time-shifting video programs. Later on, technical refinements made it possible to use the machines for virtually every stage in the production, editing and postproduction of video programs.

The Ampex VR-1000 was the first video recorder to use the revolutionary concept of moving the heads as well as the tape, the head motion being transversely across the width of a 2-inch-wide tape. In all modern recorders, transverse scanning has been replaced by a helical scanning configuration; but the basic concept remains the same. This applies not only to the professional recorders used by broadcasters, but to the vast market for home video recorders.

Over the last decade, some 50 million consumer video cassette recorders have been manufactured annually. This figure far exceeded the production of hard disk drives for personal computers, where an annual rate of 50 million units was attained only in the year 1995. Virtually every home worldwide that has a television receiver has also a video cassette recorder.

It is arguable that the worldwide social impact of Ampex's 1956 realization of the video recorder has been far greater than that of all the other concurrent technologies, including semiconductors and computers.

EARLY ROTARY-HEAD VIDEO RECORDER ATTEMPTS

Attempts to make television magnetic tape recorders with fixed heads started in the early 1950s. As described in Chapter 10, none of the many fixed-head developments progressed much further than a demonstration prototype.

In 1951 Alexander M. Poniatoff, Ampex's president, allotted $14,500 to investigate video recording, and Charles Ginsburg joined Ampex specifically to work on the first video recorder project, which started in December 1951. The key ideas were first, the use of very high head–tape speeds (to be achieved by using rotary heads) and second, the use of amplitude modulation (AM). Following suggestions made earlier by Marvin Camras of the Armour Research Foundation, Ginsburg worked on an arcuate scanner system.

There are three major classifications of rotary-head recorders: arcuate, transverse, and helical. In arcuate-scan machines, the axis of revolution of the head wheel or drum is almost orthogonal to the surface of the tape. In transverse-scan machines, the axis of revolution of the drum is parallel to the surface of the tape and parallel to the direction of motion of the tape. In helical-scan machines, the axis of revolution of the drum is parallel to the plane of the tape but almost orthogonal to the direction of motion of the tape.

In the first Ampex arcuate design, the head–tape speed was no less than 2500 in./s. The tape speed was 30 in./s with 2-inch-wide tape. The arcuate drum had only three heads and, perhaps surprisingly, used rotary capacitors. The intended bandwidth of the demodulated video recording was 2.5 MHz. In May 1952 this low-priority project was terminated, according to Ginsburg, "in favor of a crash project to turn out a one-of-a-kind instrumentation recorder." In August 1952 the project was resumed, and in October 1952 it "demonstrated an almost recognizable picture," as Ginsburg put it. At this demonstration, the president of Ampex, Alexander Poniatoff, made his famous statement: "Wonderful! Is that the horse or the cowboy?"

In March 1953 work started on a second arcuate design, which now had four heads on a drum of 1.25-inch radius with 2-inch-wide tape. This change was made so that, with each head sweeping 105 degrees of arc, it was possible to switch electronically between the two pairs of diametrically opposed heads. The head–tape speed was now reduced to 1750 in./s. An amplitude modulation (AM) scheme was used, without ac bias, again for a 2.5 MHz video bandwidth.

The principal picture defect noted in this four-head arcuate system was named "venetian blinds," and it referred to the visible picture artifacts produced by errors in merging exactly the sequential demodulated video signals from each pair of heads. In this system, the venetian blinds or "banding" arose from two major sources. First, differing head efficiencies and the head–tape spacing variations inherent in arcuate scanners caused the analog (AM) signals reproduced from the heads to be unequal. Second, geometrical errors in both the head and tape positions caused timing errors. The four-head arcuate program was stopped in June 1953, with the understanding that the arcuate-scan topology could not be controlled and that amplitude modulation (AM) was probably unsuitable.

It is interesting that Ampex's first and second attempts to produce a rotary-head video recorder were in the arcuate configuration. The immediate appeal of arcuate-scan recorders lies in the simplicity of their tape loading. Indeed, the arcuate drum need only be presented, almost orthogonally to the tape; no tape threading operations are required. Arcuate designs were not only used in the first two Ampex attempts, but also were resurrected twice more by Ampex (in the 1960s and 1970s) for proposed, fast loading, automatic video recorders for broadcasting advertising commercials. In the early 1990s, Recording Physics (San Diego) made yet another attempt at an arcuate-scan (data) recorder.

To date, all arcuate-scan recorders have failed before ever going into production. The principal reason for these failures is the extreme difficulty of maintaining satisfactory head–tape contact over the curved, arcuate, recorded track. Proper head–tape contact depends on maintaining a specific and constant head–tape contact pressure. This cannot be achieved in the arcuate or, indeed, any other curved-track topology, because the tape tension cannot be made isotropic in the plane of the tape.

THE FIRST QUADRUPLEX RECORDER (VRX-1000)

On September 1, 1954, the Ampex video recorder project was restarted in earnest. Charles Ginsburg was now joined by Charles Anderson, Ray Dolby, Shelby Henderson, Alex Maxey, and Fred Pfost, the team that finally succeeded in 1956 with the VRX-1000, the first quadruplex video recorder.

Charles Anderson was the inventor of the narrow-band frequency modulation (FM) scheme used in all video recorders until the advent of digital video recording in the late 1980s. Ray Dolby is better known today for his enormously successful company and designs in noise suppression techniques for audio recording. On the VRX-1000 project, Dolby apparently was the first to realize the advantages of transverse scanning. Shelby Henderson performed an invaluable role as the team's model maker. Alex Maxey was responsible for the topology and design of the tape transport and the (transverse-scan) head–drum assembly. Later, in 1958, Maxey was among the first to implement the helical-scanning concept, on which technology all subsequent professional and consumer helical-scan recorders are based. Fred Pfost had the task of developing and designing the video heads.

The first major change was the abandonment of the arcuate geometry and the adoption of the transverse-scan, rotary-head technique. This idea apparently came to Dolby "in the middle of the night." The 2-inch-diameter drum carried

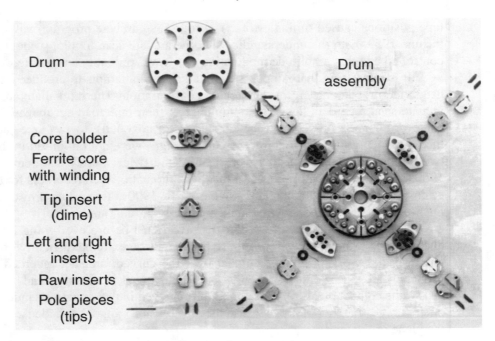

Drum ———

Drum assembly

Core holder ——

Ferrite core with winding ——

Tip insert (dime) ——

Left and right inserts ——

Raw inserts ——

Pole pieces (tips) ——

Figure 11–1   Exploded diagram of the VR-1000 head wheel, showing details of the composite video head. [*Courtesy of Ampex Archives.*]

four heads in quadrature (Fig. 11–1). Slip rings and brushes replaced the rotary capacitors. The drum rotated at 240 revolutions per second (14,400 rpm) giving a peripheral speed of 1550 in./s (88 mph). With a 2-inch-wide tape and a tape speed of 15 in./s, the recorded tracks were within 0.6 degree of being exactly transverse across the tape width. The track width was 10 mils, with 5-mil guard bands. The video heads essentially recorded over the full width of the tape, although the top and bottom edges were later erased to permit the audio and control tracks to be recorded on those regions.

There are several significant advantages of this transverse topology. First, the track is straight, and second, errors in tape transport velocity (wow and flutter) cause tracking and amplitude errors only in the playback signal, without introducing timing errors. The straight track means that the head–tape contact pressure can be held constant along the whole track, a condition that cannot be realized with a curved-track, arcuate scan. Since the ends of the almost circular arc, which defines the arcuate-scan track geometry, are not even approximately orthogonal to the tape motion, tape wow and flutter inevitably cause timing errors in the arcuate topology.

Precise control of the tape in the critical region where it passes the rotating heads was accomplished by the use of a "vacuum guide," or "female guide," which forms the tape into the required curved "canoe." To ensure postive location of the tape position immediately adjacent to the head wheel, the guide was provided with vacuum grooves just upstream of the head wheel.

Initially work recommenced using AM in conjunction with an automatic gain control (AGC) system. While the AGC could follow relatively slow variations in the amplitude-modulated playback signal (rf) due to variations from head to head, it was found to be almost useless against short period variations in the radio frequency caused by both tape defects and momentary head–tape spacing changes. In hindsight, it will be realized that without ac bias, the AM waveform must also have suffered extreme nonlinear amplitude distortion.

In December 1954 it was decided to abandon the AM system and replace it with a "vestigial sideband" frequency modulation scheme. The great novelty, at the time, of Anderson's FM scheme was that the modulating frequency (of the video) was comparable to the carrier frequency. In normal broadcast FM audio, for example, the modulating frequency of the audio, (say 20 kHz) is much smaller than the FM carrier frequency (typically 100 MHz).

In Anderson's narrow-band FM system, the video bandwidth was to be 1.25 MHz on a 3.0 MHz carrier with a 0.5 MHz sensitivity. Sensitivity is the carrier frequency deviation for a standard 1-volt white video signal. In all FM systems, there exist upper and lower sidebands, at integer multiples of the modulating (video) frequency, whose amplitudes are governed by the sensitivity. To limit the creation of lower sidebands at negative frequencies, which would cause distortion in the demodulated video, it was necessary to use very low values of the sensitivity. Accordingly, only two lower sidebands could exist. Moreover, because of the high-frequency rolloff of the video heads, it was expected that no more than one upper sideband would be reproduced and it was expected to be highly attenuated. Hence, the appellations "vestigial sideband" and "narrow band." It seems fair to say that at the time, the actual modus operandi of this narrow-band FM system was not understood properly. A mathematically complete understanding of such systems was finally provided in 1965 by Michael Felix and Harold Walsh. A modern view of a narrow-band FM system is presented later in the discussion of the VR-2000 recorder.

The first FM system was implemented by frequency modulating the (low-pass-filtered) luminance on a 46 MHz carrier. This FM signal was then heterodyned against a 49 MHz oscillator, producing the requisite 3.0 MHz carrier. On playback, the reverse process occurred so that the final FM demodulation was accomplished at 46 MHz.

First pictures were seen in early February 1955 and were judged "gratifying." By late February 1955, pictures were available using Dolby's multivibrator circuits, which permitted the video signal to be directly frequency-modulated on a 3.0 MHz carrier. This system not only greatly reduced the circuit complexity, but also gave better results. By March 1955, according to Ginsburg, "we gave a convincing demonstration" to the Ampex board of directors. The video program, that of a shipwreck, was apparently well chosen to mask both audio and video noise!

Meanwhile, important improvements in the video heads themselves were being realized. They culminated in a composite structure, designed by Pfost in 1955, which consisted of a gapped polycrystalline ferrite core onto whose face or side (i.e., not the top surface) was bonded a pair of pole pieces made of Alfenol, a binary alloy of iron and aluminum (Figure 11–1). The Alfenol pole pieces were 10 mils thick and defined the recorded track width. The two Alfenol poles were separated by a platinum gap spacer about 3 μm thick. Amazingly, the whole structure was held together only by epoxy cement until it was locked in place by the mounting screws on the head-wheel.

Alfenol, an alloy of 16% aluminum and 84% iron, has approximately one-fourth the wear rate of Permalloy (81% nickel, 19% iron). This composite design permitted the use of higher frequency rf, and the final VRX-1000 channel specifications were as follows: carrier, 4.75 MHz; sensitivity (for 1 V white video signal), 1 MHz; and a 2.5 MHz video bandwidth. Upper sidebands were thus carried to beyond 6 MHz. It is interesting to note that this (black-and-white, luminance only) 2.5 MHz video bandwidth is the same as that of the ubiquitous VHS video recorder found today in everyone's living room.

During this period, Maxey discovered that by changing the tape tension on playback it was possible to compensate for the timing errors that occur when, for example, the writing drum diameter is not exactly the same as the reading drum diameter. Achieving such compensation was rightly viewed as a critical issue, because without it, interchange between machines or, indeed, compensation for video head wear, would have been almost impossible. It was shown that the tape tension could be varied in three ways: at the reels, by moving the vacuum guide toward and away from the drum, and by varying the vacuum.

Toward the end of 1955, it was decided that the "breadboard" machine should be repackaged, and Anderson designed what was called the Mk IV console. It was also determined to plan for a surprise demonstration at the National Association of Radio and Television Broadcasters (NARTB) convention in Chicago in April 1956.

In February 1956 a demonstration was made to about 30 Ampex senior personnel. First, a program recorded about an hour beforehand was shown. Ginsburg recalled:

> We then announced that we would record a sequence and immediately play it back. We recorded for about two minutes, rewound and stopped the tape, and pushed the playback button. Completely silent up to this point, the entire group rose to its feet and shook the building with hand-clapping and shouting. The two engineers [apparently Pfost and Maxey] who had done more fighting between themselves than the rest of the crew, shook hands and slapped each other on the back with tears streaming down their faces.

Representatives of CBS, ABC, and the Canadian and British broadcasting companies were invited in the next few weeks to see similar demonstrations. Jack Mullin, of Bing Crosby Enterprises (BCE), was also invited and he said, "It is all over for us [referring to BCE's longitudinal, fixed-head video recorder]! It was a beautiful picture, better than ours!" All were sworn to secrecy. It was also decided to mount a surprise showing at the annual CBS affiliates meeting the day before the beginning of the NARTB.

Finally, the Mk IV was broken down and shipped to Chicago. Right up to the last moment development work continued. Of particular note were the almost continuous flows of improved heads from Pfost and improved tapes from William Wetzel of 3M. The final "wonder" tape used in the NARTB demonstration was of exceptional surface smoothness (for good short-wavelength recording) and exceptional toughness and durability (for long life). Du Pont supplied the base film, an experimental polyethylene terephthalate (Mylar), the plastic binder system was a polyether–urethane, and the γ-Fe_2O_3 particles of 1 µm length were transversely oriented. Such tapes were good for perhaps 100 passes.

The CBS private showing caused the CBS engineers to complain that the signal-to-noise ratio (SNR) was too low. Indeed, it seems that engineers always feel SNRs are too low. The following day, April 15, 1956, at the NARTB public demonstration, the very latest 3M tape was used. Now, the broadcasters in the audience saw what Wetzel later described as "photographic-quality pictures." There was a moment or two of stunned silence, then an outburst of cheers, stamping feet, whistles, and pandemonium!

Figure 11–2 The successful 1956 NARTB machine, with the team that produced it (left to right): Fred Pfost, Shelby Henderson, Ray Dolby, Alex Maxey, Charles Ginsburg, and Charles Anderson. [*Courtesy of Ampex Archives.*]

During the remainder of NARTB, and the two weeks following, Ampex wrote orders for about 100 machines at $45,000 each. Fourteen VRX-1000s were built, with 11 being delivered to customers, starting with CBS in Hollywood and NBC in Burbank. The first video recording went on the air on November 30, 1956, with *Douglas Edwards and the News*.

The successful team of five engineers is shown with the NARTB machine in Figure 11–2.

SUBSEQUENT DEVELOPMENTS

The VR-1000

Very shortly after the NARTB demonstration, Ampex entered in an agreement with RCA whereby, for a four-month period, a free exchange of technical ideas was permitted. Ampex's gain was access to RCA's color TV technology, with RCA acquiring Ampex's quadruplex scanner know-how.

In the years 1959–1961 Ampex and Sony engaged in a slow-motion duet, which involved many visits of engineering teams from Redwood City to Tokyo and vice versa. Ampex's main interest was access to Sony's knowledge and experience in producing solid state electronics. The negotiations apparently fell through over patent and intellectual property disputes.

In 1964 Ampex formed a 49%/51% joint venture Japanese company with Toshiba. The company, named Toamco, was to be responsible for video recorder manufacture and sales in Japan, with Ampex covering the rest of the world. With Ampex's technical assistance, Toamco progressively undertook the manufacture of the machines in Japan.

Over the years numerous improvements were made to the original quad recorder. As soon as orders began to materialize, it was realized that an improved method of assembling the video heads was required. Pfost developed a composite head using two blocks of Alfesil (5% aluminum, 10% silicon, 85% iron), which were brazed together and then sliced into 10-mil-thick pieces. Alfesil, also called Sendust, is mechanically hard, and the head life was extended from less than 100 to several hundred hours. The gap spacers used were either sputtered Al_2O_3 or SiO_2. Each head was now mounted on a very stiff beam spring, which permitted adjustment with a tapered screw to correct "quadrature" (tangential) timing errors.

The improved heads were now connected to individual slip rings, thus reducing reading head noise by almost a factor of 2. The slip-ring brushes are clearly visible in Figure 11–3, which shows the entire VR-1000 scanner assembly. This in turn permitted the introduction of what was later called "low-band

Figure 11–3 The complete VR-1000 scanner assembly, showing the head wheel, the vacuum guide, and the five slip-ring brushes. [*Courtesy of Ampex Archives.*]

color." In this Ampex system, the 3–4 MHz chrominance portion was separated from the 0–3 MHz luminance signal, decoded to I and Q components, and reencoded on a stable subcarrier. This stabilized color information was then recombined with the relatively unstable 0–3 MHz luminance signal. It is hard to visualize what this "heterodyne" color must have looked like, with stable, but basically incorrectly registered, color on unstable jittering luminance!

At that time, mechanical timing errors of several microseconds were common, and it was realized that great improvements would be realized with the introduction of timebase correction circuitry. In 1959 Intersync was introduced, followed in 1960 by Amtec, developed at CBS, and in 1961 by Colortec. Colortec reduced the timing errors to less than 5 ns and made possible the introduction of the "direct color" recording technique used in the VR-2000 and also electronic editing of color recordings.

The VR-1000 machine originally had no provision for electronic editing. In the manual editing method used, the 2-inch-wide tape was carefully slit, parallel to the recorded video tracks and at one of the control track pulse or magnetic transition positions. The new material, similarly prepared, was then butt-spliced to the old tape with a 2-inch-long piece of 0.25-inch adhesive tape. To ensure that the splice would be invisible, it was made to occur during the vertical blanking interval, that is, between the video fields. To locate the video and control tracks, the tapes had to be magnetically developed and the recordings made visible by dipping the tape into a liquid suspension of carbonyl iron particles.

By 1963 Ampex introduced Editec, which permitted electronic editing and gave producers and editors no less than frame-by-frame recording control. In this system two VR-1000 quadruplex machines are required. Even today, it is amazing that the signals from the two transports can be servo-controlled so that they both arrive, within less than 5 ns asynchronism error, at the desired editing frames. Again the actual edit is made in the vertical blanking interval and is, therefore, not visible on the TV screen. Figure 11–4 shows the president of Ampex, Alexander M. Poniatoff, and a VR-1000.

The VR-2000

In 1964 Ampex introduced the VR-2000. This machine retained the same 2-inch quadruplex head wheel and was backwardly compatible with recordings made on the VR-1000 series. Otherwise, there were major changes: for example, solid state electronics replaced the several hundred vacuum tubes, and rotary transformers replaced the venerable slip rings of the VR-1000. The video tape now had γ-Fe_2O_3 particles about 0.5 μm long, yielding higher signal-to-noise ratio. Joseph Chuppity designed new Alfesil-brazed video heads, without the ferrite core, which had considerably improved frequency response. These new heads had a higher efficiency than the Alfenol-tipped heads, mainly because the coils were placed much closer to the head–tape contact region. The "Chuppity" heads were introduced first in 1962.

By this time, a very considerable body of knowledge, both experimental and theoretical, had been accumulated at Ampex about the narrow-band FM systems used in all the quadruplex video recorders. A brief account of this understanding follows.

First, it was understood just how the demodulated, or baseband, video SNR was determined by both the properties of the tape and the characteristics of the narrow-band FM system. The video signal-to-noise power ratio was shown to be proportional to the tape particle packing density. A reduction in particle size from 1 μm to 0.5 μm causes an increase in the number of particles per unit volume by 2^3, which is almost a factor of 10, equivalent to a 10 dB increase in SNR. It was also shown that increases in track width or FM sensitivity, or a decrease in carrier

Figure 11–4   The president of Ampex, Alexander M. Poniatoff, and the top plate of a VR-1000. [*Courtesy of Ampex Archives.*]

frequency, should yield proportional increases in SNR. (The head–tape speed and video bandwidth are not considered to be variables.)

Second, it was understood how the characteristics of the narrow-band FM system controlled the distortions in video. In a brilliant analysis, Felix and Walsh showed that a specific overall transfer function of the recorder was required, and that to minimize distortion, only certain choices of the narrow-band FM parameters were acceptable. First, they demonstrated that if the overall transfer function fell in a strictly linear manner from a maximum at the lower band edge to zero at the upper band edge, then the zero-crossing positions of the frequency-modulated rf waveform were unchanged. With this "straight-line" equalization, therefore, it is possible for absolutely distortion-free video to be recovered from the zero crossing positions. In practice, the frequency demodulation to video is accomplished by hard limiting, the generation of impulses at each zero crossing, and low-pass filtering.

Also it was realized that there might exist, within the passband of the rf, spurious signals that could lead to distortions in the video. These distortions are of two kinds: "unavoidable" and "avoidable." Unavoidable distortions are inherent in the choice of the FM system parameters. Avoidable distortions arise from the less-than-perfect design or performance of components in the FM system.

The most important unavoidable distortions come from negative frequency or "folded" sidebands and from the lower sidebands of the carrier third harmonic. Avoidable distortions typically are caused by a lack of dc balance. For this reason, the entire rotating-head assembly in the VR-2000 was magnetically shielded against the earth's magnetic field.

The following FM parameters were chosen for the VR-2000 "high-band" color recorders: a carrier frequency of 7.9 MHz, a video bandwidth of 4.5 MHz, and an FM sensitivity of 2.1 MHz/V, with a straight-line passband from 1 MHz to 15 MHz. For phase-alternating line (PAL), a European standard, the parameters chosen were 7.8 MHz, 6.0 MHz, and 1.5 MHz/V, respectively. The color subcarrier in the U.S. standard, National Television Systems Committee (NTSC), is at 3.58 MHz (4.43 MHz in PAL), and the VR-2000 high-band system was thus able to record the entire composite waveform. Accordingly, this system was also called "direct color." The high-band FM system parameters were used without change in the professional helical recorders that followed the quad machines.

The video SNR of the VR-2000 high-band color recorder was 47 dB, peak-to-peak to rms. (It is customary in the television industry to quote peak-to-peak to rms SNR because it more closely suits the "peaky" nature of the video signal.) One of the many requirements of a video recorder is the ability to rerecord many times from tape to tape. This requirement arises in both program editing and distribution operations. Each generation is degraded because the incoherent tape noise power adds. Thus, after two generations the SNR is reduced 3 dB to 44 dB and, after four generations, by 6 dB, to 41 dB. More than eight generations of copies could be made before the SNR deteriorated to less than 38 dB, which is the threshold for "visible" noise.

It was found, however, that long before the deteriorating SNR became a limiting factor, the buildup of timebase errors caused intolerable picture degradation, particularly in the color signal. In the first electronic timebase corrector, the Amtec, timebase errors were detected and corrected on a video line-to-line basis by comparing a synchronizing pulse with a reference horizontal signal. The Colortec system operated by comparing the "color burst" signal against a reference color subcarrier signal. Timing-error information is thus available only in the horizontal intervals, and timebase corrections were made only during those intervals.

Major steps forward were made by Charles Coleman, who had designed both the Amtec at CBS in 1960 and the Colortec at Ampex in 1961. In 1966 he invented the "velocity compensator," which took the timebase error information derived as in Colortec, but applied the correction uniformly over the duration of the horizontal line. The assumption was that the origin of the timebase error was simply that the playback scanning speed was not identical to the writing scanning speed.

The results for the velocity compensator were dramatic. In 1966 Coleman gave a paper at the National Association of Broadcasters meeting. During the presentation he ran a tape showing a master recording followed by 10 generations of rerecording. Following the sixth generation and on the way to the tenth, the audience rose to its feet and gave him an unprecedented round of applause. Shades of 1956!.

RELATED DEVELOPMENTS

Before this account of the quadruplex story ends, three related Ampex developments deserve mention: the Videofile, the Tera-Bit Memory, and the Digital Cassette Recording System. These systems are not video recording systems, therefore it would be inappropriate to describe them in detail here. The following short accounts are justified, however, because all these systems were based on the use of quadruplex recording technology.

The Videofile

The Videofile, developed in the late 1960s, was a document image storage and retrieval system derived directly from the VR-2000. The system allowed for the storage of 170,000 pages or documents on a standard 12-inch-diameter (4800 feet) reel of 2-inch-wide video tape. The average search time to a document was 50 seconds, using a 380 in./s tape search speed, which scanned 1140 pages per second. The document images were in the so-called portrait mode (8.5 inches wide × 11 inches high) with 1280 lines of horizontal resolution (higher than HDTV) and 1240 active scan lines. Videofile transports were operated under the control of a Data General PDP-11 computer.

Eight Videofile systems were made. The first customer was the Southern Pacific Railroad, which used it for the daily "roundup" of its freight cars. Several insurance companies and hospitals ordered Videofile systems. The last customer, and perhaps the most well known, Scotland Yard in the United Kingdom, used its system for the world's first "online" fingerprint file.

The Tera-Bit Memory

In 1970 Ampex developed a digital mass memory system called the Tera-Bit Memory. The TBM, at the time, was by orders of magnitude the largest capacity digital storage and retrieval system in the world. It was also based on the quadruplex VR-2000 technology. TBM used 2-inch-wide tape and special 2-inch-diameter head-wheels with eight heads, placed every 45 degrees around the wheel. The drum speed was reduced to give a head–tape speed of about 800 in./s. Digital data recording occurred at 7500 b/in. by means of a VR-2000-like high-band FM system. The many heads allowed "spatial redundancy" to be used to achieve a low bit-error rate. Twenty tape transports were required to store one terabit (10^{12} bits) of data. The largest TBM system assembled had 36 tape transports for a total capacity of 1.8×10^{12} bits of data.

Customers of TBM were exclusively in the intelligence community and in other government agencies. Fewer than six systems were made.

The Digital Cassette Recording System (DCRS)

The DCRS, introduced by Ampex in 1983, is both the most recently designed and the only transverse-scan recorder in production today. It is probably the last of the breed. The DCRS uses 1-inch-wide tape with cobalt-modified γ-Fe_2O_3 particles about 0.25 μm long. A special tape cassette holds 1200 feet of tape. The transverse-scan head wheel is 40 mm in diameter. The original DCRS had six heads (at 60-degree spacings); later versions have a dozen heads (at 30-degree spacings). Rotating at 30,000 rpm, the recorder has a head–tape speed of 2560 in./s, which is the highest ever attained in any magnetic recorder. The written track width (1.9 mils) is wider than the track pitch (1.73 mils). The linear density is 46,000 b/in. The heads are made of conventional polycrystalline manganese–zinc ferrite, with $\pm$15-degree gap azimuth angles between adjacent heads. The cassette holds 3.8×10^{11} bits of user data, and a bit-error rate of 10^{-10} is achieved.

Transverse scanning was chosen for several reasons. First, the recorders were intended to operate in adverse environmental conditions. In transverse scan, errors in tape position due to tape wow and flutter cause tracking errors only. Second, the machines were intended to operate in an incremental mode, in which a machine could change from standby (tape disengaged and stationary, but head wheel rotating) to writing mode in less than 1 second. The simple tape topology of transverse scan greatly facilitates such rapid tape acceleration and head engagement. The system is fully addressable and performs exactly like an electronically searchable solid state memory of capacity 3.8×10^{11} bits.

DCRS recorders are the most advanced digital instrumentation machines available today. They are widely used in testing modern aircraft such as the F-16 and the Boeing 777 and are installed in the space shuttles and the Mir space station.

THE DEMISE OF QUADRUPLEX VIDEO RECORDERS

Quadruplex video recorders were constantly refined and developed during the 1960s and early 1970s. During this period, they became able to fulfill every operational requirement of TV program producers, editors, and distributors. Despite their technical success, however, quadruplex recorders were supplanted, during the 1970s, by helical-scan machines. By 1978 Type C helical machines, such as the Ampex VPR-2, became the industry standard for video recording. Although helical-scan recorders are treated specifically in the following chapters, it is appropriate to add a few comments here on the question of quadruplex versus helical scanning.

Alex Maxey introduced the concept of helical scanning at Ampex in 1958, and the first production machine, the VR 1500, appeared in 1961. The immediate appeal of helical machines is that the length of the individual recorded track can be made large enough to record a complete TV field. This eliminates the "venetian blind" type of picture defect, because a TV field contains all the odd (or even) numbered horizontal TV scan lines.

In quadruplex machines, the length of each individual recorded track is somewhat less than the 2-inch width of the video tape, and such systems could record only 16 to 17 horizontal TV lines. Unless great care is taken when switching from one video head to the next, slight analog differences in amplitude and timing give rise to picture impairments of the venetian blind, or banding, type. Since there are 525/2, or 262.5, horizontal lines per field in NTSC, 16 bands appear in the TV image. During the development and refinement of quadruplex recorders, banding was rendered almost imperceptible, but this advance was achieved only at the expense of a great deal of complex electronics (e.g., Colortec, velocity compensation).

In the 1960s, even though a helical-scan machine would not suffer from banding, the machines were essentially unusable as TV recorders. This was basically because it was not possible to construct either analog or digital delay lines of sufficient capacity (1 field) to perform the necessary timebase correction. In the early 1970s, however, advances in large-scale integration of silicon devices made possible, for the first time, the economical design of a digital timebase corrector having one-field capacity.

Thus the question became which is better: a single large (about 1.5 Mb) digital timebase corrector, or four sets of complex analog correction circuitry? In a nutshell, helical displaced quadruplex because it became a less expensive, more easily maintainable way to produce TV images free of banding.

In 1977 another Ampex invention further assured the demise of quad video recorders. This was called Automatic Scan Tracking (AST). By mounting the video heads on a piezoelectric actuator, it became possible to track accurately prerecorded video at any tape speed from 100% forward to a complete stop and even about 20% reverse. AST made it possible to perform all the stop and slow-motion effects, so popular in sports programming, right on the helical-scan video recorder itself. Previously, an auxiliary piece of equipment, the Ampex HS-100 "stop-motion" thin-film disc recorder, had been required.

The demise of the quadruplex machines was now a certainty!

CONCLUSIONS

It is difficult, today, to recall the near perfection of a well-adjusted analog FM video recording. This recall is made more difficult by the constant publicity erroneously suggesting that today's digital TV is of higher resolution or "superior digital clarity." The current, full bit rate, professional digital recorder formats have the same resolution but slightly higher signal-to-noise than the FM analog recorders. Appreciable bit rate compression such as occurs in MPEG-2 (the current standard of the Motion Picture Experts Group), of course, reduces both the resolution and SNR.

Most of the elegant analog technology was continued in the professional FM helical-scan video recorders like the Type C. All this sophisticated FM technology was abandoned by the late 1980s, however, with the advent of digital video recorders such as the D-1, D-2, and D-3. Surely, historians will judge this change as a triumph of brute force over elegance.

Analog recorders gave way to digital machines for the same two basic reasons that apply in the cases of other electronic devices. First, proper performance of an analog recorder requires the maintenance of extremely precise adjustments. Second, analog errors can build up without limit in multiple generations or dubs. In digital recorders and digital circuitry, such exact adjustments are not required, and errors can be corrected in each generation of copies.

ACKNOWLEDGMENTS

The writer joined Ampex in 1962, six years after the unveiling of the first Quadruplex recorder, and this chapter could not have been written without help from others. In particular, I acknowledge the assistance rendered by Charles

Coleman, Beverley Gooch, Jerry Miller, and Kurt Wallace in proofreading this chapter. I would also like to acknowledge the countless hours of explanations of technical matters so freely provided by so many talented people during the 22 years I was at Ampex. For example, I believe that I learned more about how a high-density, analog FM channel really works from Michael Felix than from any other individual.

REFERENCES

Felix, M. O., and H. Walsh, "FM System of Exceptional Bandwidth," *Proceedings of the IEEE*, **112**, 9 (1965).

Ginsburg, C. P., "The Horse or the Cowboy: Getting Television on Tape," *Journal of the Royal Television Society*, November/December 1981.

Ginsburg, C. P., "The First VTR: A Historical Perspective," *Broadcast Engineering*, May 1984.

Mallinson, J. C., "The Signal-to-Noise Ratio of a Frequency-Modulated Video Recorder," *European Broadcast Union (EBU) Review*, **153** (1975).

Mallinson, J. C., *The Foundations of Magnetic Recording*, Academic Press, San Diego, CA, 1993.

Videotape Recording, Ampex Corporation, Redwood City, CA, 1986.

12 Helical-Scan Recorders for Broadcasting

Hiroshi Sugaya

VIDEO RECORDERS FOR EDUCATIONAL AND INSTITUTIONAL USE

After Ampex developed the quadruplex video tape recorder (VR-1000) in 1956, the main television broadcasting stations in the United States, including the Public Broadcasting Service, started to use it. For many smaller television stations such as local television or community antenna television, the quad recorders were a substantial investment. Both operators of these stations and institutional users of television would have been very happy to have a less expensive video tape recorder, even if its timebase stability and picture quality were not as good as those of the quad machines. At this time, transistors became available as a market commodity. By extensive use of transistor circuitry, engineers could design a video tape recorder that was more compact and less expensive.

The first commercially available video tape recorder for institutional television applications was the Ampex VR-660, produced in 1961 (Fig. 12–1). Using two heads helically scanning a 2-inch tape, the VR-660 could record one field of a picture signal, including audio and control signals, on one scanning line. This made the circuitry relatively simple. This machine could record over 4 hours of programs on a 12-inch reel, but since a double-size head drum was required, it was very bulky. Soon after, in 1962, Machtronics developed a smaller, 1-inch tape recorder (MVR-10), followed by the Precision Instruments Model PI-3V in 1963. The tape width and speed were just half those of the Ampex VR-1000.

It is physically impossible to record a complete video field on a single helical track using the omega-wrap configuration. In 1962 Sony developed a 2-inch video recorder (PV-100) using a new concept of recording, the so-called 1.5-head

Figure 12–1   The first 2-inch, helical-scan video recorder, the Ampex VR-660. [*Courtesy of Ampex Corporation.*]

system (Fig. 12–2). At that time, video head technology was far from mature, and it was difficult to make two heads having the same characteristics. Consequently, two successive scans did not look the same. Recording with one head presented problems in dealing with the audio and control tracks. The 1.5-head system treated the problem by splitting the television signal so that one head recorded the video portion and a second, the "half-head," recorded the vertical synchronization part. Figure 12–3 shows the omega-wrapped tape path as it is scanned by the

Figure 12–2   The world's first 1.5 head recorder, the Sony PV-100. [*Courtesy of Sony Corporation.*]

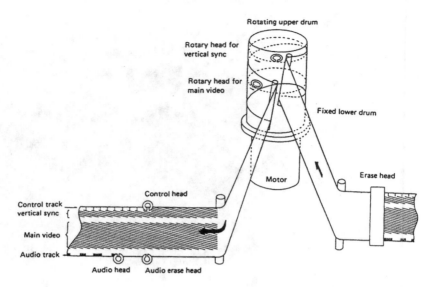

Figure 12–3   Principle of the 1.5 head system. [*Courtesy of McGraw-Hill and IEEE Press.*]

heads and the final layout of the recorded signals, including audio and control. These recorders were much smaller than the Ampex quad machine, but still required two people to transport them.

Meanwhile, Ampex developed the VR-7000 series tape recorders. These machines achieved a very smoothly running head-to-tape interface by using an air–film helical-scan recording method. With this interface, slow-motion sequences could be played back up to 20 times without degrading the high-quality color picture. This video recorder series also had good machine-to-machine interchangeability, but recording short wavelengths was still a major technical problem.

A problem of a different sort was encountered in making a video recorder for both the NTSC and PAL/SECAM areas, with their different power line frequencies of 60 and 50 Hz. (These U.S. and European standards are discussed in Chapter 9.) Because the head drum rotates synchronously with the power line frequency, the drum radius would have to change accordingly to maintain the same head-to-tape linear velocity. An engineer at Ampex had a very good idea and established Westel Company (later Echo Science) to implement it. The secret was to divide the frame into segments and change the number of segments in one recorded field according to the synchronized frame frequency (50/60 Hz). This so-called segmented (sequential) system was used in several generations of video systems, even after the fully digital D-1 system described in Chapter 14 was developed. The disadvantage of this recorder was the degradation of picture quality caused by the segmenting: five elements for 60 Hz (NTSC) and six elements for 50 Hz (PAL/SECAM). Banding occurred in the picture because of differences between the two heads. The electronic circuitry was complex, but the machine was small and of course had the flexibility of operating with different television systems. The primary users of this video tape recorder were the U.S. armed services.

One of the more successful institutional video recorders in the United States was the International Video Corporation's Model 800, developed in 1967. This recorder achieved high performance and compact size using one hot-pressed ferrite (HPF) head on a drum. Because recording took place with only one head, the picture quality was very constant. The developers were faced with the problem that, in the α-wrapped, one-head, helical-scanning recorder, there was no space to record audio and control signals. The problem was solved by recording the audio and control signals on the video track before the video signal was recorded. This idea of overwrite recording was not new at that time, but IVC was the first to use it on a commercial video tape recorder. At that time, the number of television broadcast stations mushroomed, but there were no inexpensive satellite stations supplying a variety of programs for them. The Model 800 video recorder was a good match for IVC's requirement for a high-quality but inexpensive recorder to supply the stations' programs.

THE BEGINNING OF COMPETITION FOR THE AMPEX QUAD RECORDER

Stimulated by the success of the Model 800, IVC engineers thought that there was a good opportunity to make a true broadcast-quality video recorder for the big, key stations. The Ampex quad video recorder (high-band color machine) was adequately meeting requirements, but magnetic recording technologies were improving almost every year. Broadcasters, not only in the United States but also in Europe, expected further improvements in performance (e.g., absence of moiré patterns, lower tape cost). The first quad video recorder could record a wavelength no shorter than 5 μm, but by 1970 wavelengths down to 3 μm could be recorded with adequate signal-to-noise ratio. European broadcasters especially required better picture quality, because of the wider video band and difficult modulation system of PAL/SECAM.

IVC engineers designed the Model 9000 video recorder to cut the tape consumption (and tape cost) to just half that of the quad recorder, while maintaining the picture quality (Fig. 12–4). However, the Model 9000 still used 2-inch-wide tape, which turned out to be a major error. Using the segmented recording system, this machine could record PAL/SECAM signals as well as NTSC without change of tape speed (8 in./s) and head–drum rotating speed. The recorder was unveiled during the 1971 Television Symposium in Montreux. It did not have the weak points of the quad recorder in Europe such as moiré, banding, and the need for two audio tracks (cue and address).

Meanwhile, Ampex engineers developed the "super-high-band system" to improve picture quality by upping the modulation frequency to eliminate the

Figure 12–4   The IVC 9000.

moiré patterns caused by interaction between the video signal and modulation sidebands. Broadcasters could add this feature to their existing quad recorders relatively inexpensively by changing the modulation circuits and head assembly units. There was little incentive to pay the price of buying the newly developed IVC recorder, even though the tape cost was cut in half. As it was, IVC managed to sell 300 Model 9000 recorders, but was forced to declare bankruptcy in 1977. If IVC had reduced the tape cost to one-third, broadcasters might have been more interested.

In the mid-1970s, Fernseh and Philips undertook a joint project that resulted in the BCR recorder, the first helical-scan machine not only designed to meet the specifications of the European Broadcasting Union (EBU) but directed specifically to broadcast applications. Unfortunately, design faults eventually forced Fernseh to abandon the project after spending some 40 million deutsche marks. Later, Bosch–Fernseh developed Type BCN (Fig. 12–5), a new 1-inch tape recorder following the Society of Motion Picture and Television Engineers (SMPTE) Type B standards, formulated using the segmented method for PAL/SECAM. This recorder did have only one-third the tape consumption of the quad recorder. The Bosch–Fernseh recorder competed successfully against the Ampex quad recorder, but primarily in Europe.

The bulky tape of the quad recorder caused many problems, such as requiring a large space for storage. NHK (Japan Broadcast Corporation), for instance,

Figure 12–5 The SMPTE Type B recorder, Bosch–Fernseh Model. [*Courtesy of Bosch–Fernseh.*]

designed a big tape warehouse with a fully automatic belt–conveyer system. A machine selected a tape box and brought it to the room where the video recorder was located. The tape box was put on a large rotary table, and waited its turn to be used. All remaining operations, like tape transport and threading, were done by hand. With the inclusion of storage and handling, the cost of tape was very high. What many broadcasters wanted was further reductions in tape consumption, small tape reels for easy handling, and better picture quality.

THE FIRST HELICAL-SCAN RECORDER TO REPLACE THE QUAD RECORDER

Initially, the manufacturers determined all newly developed video recorder specifications. By 1972, there were already several 1-inch tape video recorders in the world market, manufactured by such companies as Ampex, Bosch, IVC, and Philips. American and European broadcasters struggled to define a second-generation video recorder, based on advanced magnetic recording technologies, such as high-coercivity tape and ferrite heads, which would equal or exceed the quad in picture quality. A white paper laying out a "straw man" format was proposed by two major networks (CBS and ABC) at the 1977 SMPTE winter conference. The main points were as follows: 1-inch tape; a helical-scan, nonsegmented system; and a complete video signal, including vertical synchronizing signals and two-plus-one audio tracks. Of the several 1-inch helical video recorders on the market, the Ampex VPR-1 and the Sony BVH-1000 were closest to the white paper's requirements.

The SMPTE formed a working group to examine the two specifications and to attempt to find agreement between Sony and Ampex on a common set of specifications. The major technical differences needing resolution were the design of a method to record the vertical synchronizing signal, the diameter of the video head drum, and the longitudinal position of the audio tracks relative to the video signals. The working group met frequently to work out compromise specifications. For example, the head drum diameter of Ampex was 134.620 mm, whereas that of Sony was 135.000 mm, a difference of only 0.3%! Finally, Sony adopted the size of the Ampex drum, and Ampex adopted the basic principle of the 1.5-head system. The latter could compensate for the deficiencies of the one-head helical method and record a perfect video signal. A year later, the two companies would merge their specifications into the SMPTE Type C format (Fig. 12–6).

Figure 12–6 The SMPTE Type C recorder, Sony Model. [*Courtesy of Sony Corporation.*]

Another working group also defined the Bosch–Fernseh BCN format as the SMPTE Type B format.

A BROADCAST VIDEO RECORDER FOR ELECTRONIC NEWS GATHERING

The first portable broadcast video recorder, the Ampex VR-3000 portable quad machine, allowed a single (large, strong!) person to carry the recorder on his back and to operate the camera as well. Usually, however, the recorder was placed on a small wheeled cart like those used for today's carry-on airplane luggage. The VR-3000 was used quite widely for sports and news recording.

In 1969 Sony, Panasonic (Matsushita Electric Industrial Company) and JVC (Japan Victor Company) jointly developed the U-Format video cassette recorder for consumer purposes. These companies put the cassette recorder on the Japanese market. The machine was, however, heavy and expensive for home use. One hour of tape cost nearly $100. When Sony tried to market this machine in the U.S. institutional market, large companies like IBM adopted it for use in training. In 1974 CBS asked Sony to modify the machine to replace 16 mm film machines for news gathering. Thus, the Sony Broadcast Video Series was born in 1976 (Fig. 12–7). With one person operating a camera and a second person carrying the recorder, a news team could move quickly to any

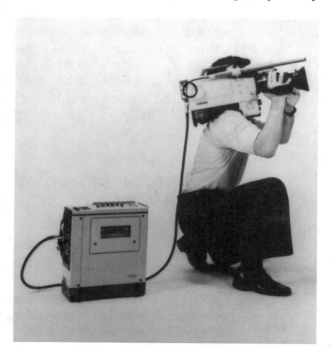

Figure 12–7 The world's first video cassette recorder for Electronic News Gathering. [*Courtesy of Sony Corporation.*]

location where a story was developing. This system was known as ENG (electronic news gathering).

To use this kind of video cassette recorder with its poor time base stability in the broadcast field, a digital timebase corrector (TBC) was indispensable. After the development of high-capacity, dynamic random access memory (DRAM) devices and digital technologies, Consolidated Video Systems in the United States developed a digital timebase corrector in 1973. The timing was good for adoption of this corrector for electronic news gathering. Timebase correction facilitated broadcast use of the U-format, assuring the success of the professional video cassette recorder.

Although the U-Matic ENG was a good machine, its operation was still a two-person job. There was a big demand for a shoulder-mounted camera/video system that would allow a single operator to move as quickly as he or she wanted. Panasonic and RCA pursued the development of such a product. For reasons of cost and ease of operation, they decided to use a cassette recorder and the VHS system. This was the world's first attempt at two-track recording technology on helically scanned tape. Many helical-scan recorders had been developed, but they all were using only one track to record directly NTSC or PAL/SECAM video signals. Achieving much better color picture quality involved treating the two color video component signals separately (i.e., the Y signal as the black-and-white or luminance picture signal, and the I and Q, or $R - Y$ and $B - Y$, signals as the color or chrominance signals). The development engineers tried to record two tracks using four heads instead of two heads, and recorded luminance (Y) and chrominance (I and Q) signals separately. As a large potential customer, NHK contributed to the development of this, the world's first camera–video cassette recorder. This machine, named Hawkeye, was unveiled at the Television Symposium in Montreux. Many broadcasters and manufacturers, including Sony, were surprised to see such a compact machine that could be handled easily by one operator. This new format was called Type M by the SMPTE.

COMPETITION IN THE BROADCAST VIDEO RECORDER INDUSTRY

The Type M recorder had a weak point in its color system. Although advanced digital technologies and DRAM were available at the time of development, they were expensive and not easy to obtain. To meet the delivery time, Panasonic engineers decided to use analog technologies to record I and Q color signals by means of frequency modulation. The development team also considered the use of digital technologies to record the color $R - Y$ and $B - Y$ signals after digitally compressing them by means of DRAMs, but this approach was not commercially feasible until two years later.

The Sony group immediately tried to surpass the Type M machine—they knew that digital color recording should be better than analog. The very next year they used a commercial tactic to announce their new machine using a Beta cassette as "Betacam." It took almost two years for the Betacam, designated Type L format by the SMPTE, to become commercially available. From the specification point of view, Betacam did surpass Type M on some points, such as digital color recording and smaller cassette. As usual, the latecomer improved the deficiencies of the front runner, just as in the Betamax versus VHS story. After the announcement of Betacam, the innovative Type-M machines could not keep a large market share and finally succumbed to Type L. Among analog VCR products, Betacam is one of the most popular ENG systems in the world (Fig. 12–8).

Panasonic improved its color recording method following the same approach as Betacam, also adopting metal pigment tape, which had nearly twice the magnetic performance of conventional ferric oxide tape. Broadcasters are generally very slow to adopt new things, and they did not like to use the new metal tape. This is why the first digital broadcast cassette recorder (D-1) used ferric oxide tape in spite of the very large cassette needed. After a number of substantial improvements, a new Type M format was announced as M-II. The cassette configuration was also slightly changed (a VHS cassette cannot be used on an M-II machine, and vice versa). The metal tape used on M-II recorders made for better picture quality than was available from Betacam. Sony then followed suit and improved the specification for Betacam by using metal tape in the Betacam SP. These competitions were pursued vigorously.

Figure 12–8 The Sony Betacam. [*Courtesy of Sony Corporation.*]

A NEW CONCEPT: AUTOMATIC CASSETTE RECORDERS

Commercial broadcast television systems are dominant in the United States, where the networks generate large amounts of money from commercial advertisements. Such programming is therefore very important, not only in terms of production but also in the way it is transmitted to the appropriate audience. Most commercials are less than 5 minutes long. If these programs could be put on cassette tapes, television broadcast stations could use them whenever they wanted, even changing the contents of the cassette according to the market reaction to the content.

For this application, Ampex in 1970 developed the video cassette recorder (ACR-25) for its quad system. The ACR-25 could keep 25 cassettes in the machine, and cassette selection and playback were computer-controlled. While one cassette was being played back, the next cassette was loaded on the machine. RCA also designed and sold the TCR-1 cartridge machine in the same period. Each machine had its own format. The SMPTE tried unsuccessfully to achieve a common tape format, although some specifications, such as leader content, were eventually standardized.

As time passed, the major television stations developed a strong need for computer control to eliminate human errors and save costs. The video tape recording technology evolved to meet these requirements. In the arrangement

Figure 12–9   The Panasonic MARC automated cassette system. [*Courtesy of Matsushita Electric and Industrial Company.*]

known as the "video cart machine," a robot selected a cassette and loaded it on a machine. There were four or five machines in a cart, and each waited its turn. In 1987 the Panasonic MARC (**M**-II **a**utomated **r**ecording/playback **c**assette system) provided on-air programming with preview. The machine accommodated a maximum of 1179 cassettes for extremely high programming flexibility. The MARC (Fig. 12–9) is especially effective when digital broadcast video cassette recorder systems are used.

THE DIGITAL VIDEO ERA

After the rapid development of semiconductor devices, especially very large capacity DRAMs, advanced digital technologies have moved from dream to reality. One silicon chip can contain over 10 million transistors, which can be configured into a very big digital system. Magnetic recording technologies also advanced faster than expected. By combining these technologies, very small digital video cassette recorders can be made. One example is the DVC (digital video cassette). The DVC-Pro, a modified version of DVC, was marketed as a handy camera/video ENG unit in 1997. Electronic news gathering involves obtaining many program sources from all over the world. The materials are edited for a story and reedited for a news program. Several duplications are required for each editing pass. Digital recording is essential for maintaining picture quality after many duplications. It must also be implemented in video cart machines requiring many duplications for their operation. Digital video will be the main system in broadcast and related areas. A more detailed description of digital video is given in Chapter 14.

REFERENCES

Livingstone, P., et al., "The M.A.R.C. 11 System: A Modular Multiple Robotic Record/Play Video Cassette System," *J. SMPTE,* **99,** 452–458 (1990).

Sugaya, H., "Newly Developed Hot-Pressed Ferrite Head," *IEEE Transactions on Magnetics,* **MAG-4**, 295–301 (1968).

Sugaya, H., "Video Recording," Chapter 5 in *Magnetic Recording Handbook*, C. D. Mee and E. D. Daniel, Eds., McGraw-Hill, New York, 1990.

Sugaya, H., "The Past Quarter-Century and the Next Decade of Video Tape Recording," *J. SMPTE*, **101**, No. 1, 10–13 (1992).

Sugaya, H., "Analog Video Recording," Chapter 5 in *Magnetic Storage Handbook*, McGraw-Hill, New York, and IEEE Press, Piscataway, NJ, 1996.

13 Consumer Video Recorders

Hiroshi Sugaya

THE CHALLENGE OF FIXED-HEAD CONSUMER VIDEO RECORDERS

After Ampex developed the quadruplex video tape recorder in 1956, consumers who enjoyed watching their favorite television programs started to dream of having their own "home video recorders." There were many programs people wanted to watch but could not because of their schedules. There were also many programs people wished to see again, such as favorite movies. From 1957 on, the Ampex quad video recorders were imported into every key station in Japan, including the Japan Broadcast Corporation (NHK). Sumo wrestling is one of the most popular sports in Japan, and the nature of the sport and its popularity contributed to the rapid adaptation to video tape recorder technology by the Japanese. A sumo match is usually very short, involving incredibly fast moves triggered by quick reflexes of the large, but graceful competitors. Many matches last only seconds.

NHK broadcast stations used video tape recorders in their coverage of sumo tournaments, quickly rewinding the tape after each match to replay the contest. The station urged viewers to "Look at the match by video tape recorder again!" This instant playback was really fantastic for Japanese audiences. The phrase "video tape recorder," or VTR, also became very popular in Japan.

Most mechanical engineers were wary of introducing a rotating-head system into the home environment. In other words, a fixed-head video tape recorder was the paradigm engineers were working toward. Thus as described in Chapter 12, the initial attempts to develop video tape recorders for the home were based on the use of fixed heads. All these attempts were unsuccessful, although many received considerable publicity at the time of launch and spurred increasing interest in the concept of a practical home video recorder.

THE DEVELOPMENT OF HELICALLY SCANNED HOME VIDEO RECORDERS

It turned out that even for the home market, helically scanned, rotating-head systems were the right approach. Sony developed a small video tape recorder (CV-2000) using 0.5-inch tape in 1964–1965 and marketed it for about $2000 (Fig. 13–1). This machine used a field-skip method, recording only one field in each frame. The recorded area required for a one-hour program was thus the same as that of a normal half-hour recording. Using two heads to play back the same field picture twice gave a poorer picture quality than normal recording with respect to the resolving power of vertical lines and vertically moving pictures, and led to a staccato type of distortion. Nevertheless, this recording machine, which used a 7-inch reel, the size needed for an hour's audio tape recording, was attractive because of its small size.

In 1966 Panasonic (Matsushita) developed its NV-1010 video tape recorder. This machine was the same size as its Sony counterpart and recorded for 40 minutes using a normal recording method. The picture quality was improved, but the short recording time remained a big problem. Recording time was the key issue for home video tape recorders, but until the successful development of the VHS video cassette recorder, nobody really knew the right recording time for consumer use.

Figure 13–1 The first 0.5-inch tape rotary-head B/W video recorder, the Sony CV-2000. [*Courtesy of Sony Corporation.*]

Some engineers at Panasonic came up with a great idea, which could easily increase the recording time. This was the so-called *azimuth recording method*. As shown in Figure 13–2**a**, the video head gap azimuth angle was tilted alternately by 30 degrees (for this model), and the heads recorded video tracks with no guard band. The recorded video signal was frequency-modulated onto a higher carrier frequency, which had very short wavelength. The azimuth loss on playback was very high at such a short wavelength. Therefore, during playback, the head that recorded track A picked up only a negligible amount of the signal from the adjacent track B even if it overlapped track B somewhat. Until the development of azimuth recording, the guard band that had been required to avoid pickup from adjacent tracks (Fig. 13–2**b**) was about half the video trackwidth. Azimuth recording saved 50% of tape consumption immediately. What is more, tape interchangeability was also improved with this method, because the adjacent track B served as a guard band for track A. Thus the effective guard band became about twice as effective as that of the conventional helical-scan recording method. In 1968 the world's first azimuth recording video tape recorder, Panasonic Model NV-2320, came on the market. It could record a 90-minute program with essentially the same electronics and mechanism as the NV-1010 video tape recorder.

There is an historical irony associated with this product. At the time of its development, the video manufacturers in Japan were discussing a common standard for the educational market. The elected chairman of the technical committee was from Panasonic. He made his best effort to establish the Electronic Industries Association of Japan (EIA-J) Type I standard, which used 0.5-inch tape, ran at 7.5 in./s, and used a conventional method to record one hour of black-and-white (B/W) program on 7-inch reels. He decided to discontinue the valuable azimuth recording machine NV-2320 to preserve harmony in the use of the EIA-J Type I video tape recorder!

At almost the same time, Philips developed a 10 kg "portable" B/W home video tape recorder (LDL-1000) using newly developed chromium dioxide tape capable of recording shorter wavelengths than conventional ferric oxide tape. The machine could record 45 minutes on 450 meters of 0.5-inch-wide tape.

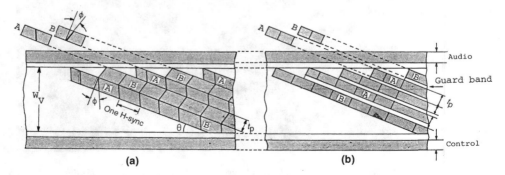

Figure 13–2   (**a**) Azimuth recording method and (**b**) guard band recording method.

A little later, Sony, Panasonic, and JVC (Japan Victor Company) discussed color video tape cassette recorders for home use using chromium dioxide tape. The azimuth recording of the NV-2320 recorder was mentioned, but not discussed further. At this time, the evaluation of azimuth recording was not extensive, especially for color recording. Use of azimuth recording in a regular format for video tape recorders had to wait until the Sony Betamax was introduced in 1975.

The idea for this home video cassette recorder (VCR) was established in 1969. According to the conventional technology of the day, it was just engineering common sense that a 7-inch reel of tape was required for one hour of recording. A cassette using two such reels would be far too bulky for home use. The engineers reduced cassette size in two ways. One was to increase the tape width from 0.5 inch to 0.75 inch. The second was to use chromium dioxide tape, which had twice the magnetic performance. Thus, the new cassette had a size of about 5.5 × 8 inches, although the thickness was increased. This new video cassette recorder, called the U-Format, was put on the market in 1971 (Fig. 13–3). The basic design of the U-Format was excellent, but the device itself was still too bulky and expensive for general home use. These first U-Format recorders were mainly used for industrial and education purposes; later on, their application expanded to broadcast electronic news gathering (ENG), as described in Chapter 12. A miniaturized version of the U-Format, incorporating azimuth recording, became the Betamax in 1975.

At the beginning of 1970, however, the mainstream of video tape recorders was still the open-reel EIA-J Type I. When this video recorder was first introduced on the educational market by several manufacturers, users requested a test of machine-to-machine interchangeability. The manufacturers hesitated in the beginning, but finally they tested every machine for interchangeability. The results were excellent and promoted very strong confidence among the users. In those days, the

Figure 13–3 The first video cassette recorder, the U-Format Sony U-Matic. [*Courtesy of Sony Corporation.*]

recorded video track was not a straight line, but bent like a sine curve. Even now, the video track is slightly nonlinear, but is controlled within a certain solid straight line. EIA-J decided to make a standard alignment tape recorder, and to supply standard alignment tapes to maintain interchangeability among all manufacturers. The alignment recorder was difficult to make, but Panasonic succeeded in 1972, producing a machine that kept the line of the video track straight within 10 μm after long-time use. In 1976, when VHS was introduced in Japan, Panasonic was still making this standard alignment tape recorder and supplied some copies of it to the companies that adopted the VHS format for secondary standard machines. Nowadays, we can easily measure video track linearity, and manufacturers can control their tape format without needing standard alignment recorders.

Nevertheless, machine-to-machine interchangeability is a kind of miracle. The VHS 6-hour-mode machine records five tracks on the thickness of a dollar bill, and the machines of different manufacturers play the cassettes back with no problem. Even now, mechanical engineers claim that the interchangeability of VHS 6-hour-mode cassettes is impossible, given the mechanical tolerances on all the parts. "Fear is often greater than the danger itself." Track width is becoming narrower and narrower for every new format. Many mechanical engineers are still warning about interchangeability every time a new advance is made. Yet, interchangeability remains little problem, even after some years.

CASSETTE (TWO REELS) OR CARTRIDGE (ONE REEL)?

To ensure proper use of a video tape recorder, the tape should be housed in a container that protects the delicate medium from dust and fingerprints. A container is also useful in positioning the tape and loading it on a machine having the complicated tape-threading path of a rotary-head system.

Many manufacturers tried to develop suitable containers. There was already a long history of containers for audio. The main difference between audio and video is the use of a rotating-head drum instead of fixed heads. The tape must be wound helically on the head drum (Fig. 13–4). From this helical threading point of view, a single-reel cartridge is one of the easiest approaches. Leader tape can be used to negotiate any complicated tape angle, and the video portion will wind up on the other reel located in the machine. One example of this cartridge was the EIA-J Type I cartridge, introduced in 1972. The biggest disadvantage of a cartridge was that it could not be removed until the tape had been wound back into the reel, which could take a few minutes if the tape was at the end of the reel. Yet 3 minutes is one of the sensory limit times of humans, according to a psychological study. Consequently, many engineers tried to use two reels in one container, either in tandem or in parallel, just like the audio compact cassette. Helical winding onto a rotating-head drum was, however, a big problem. If two tape reels are

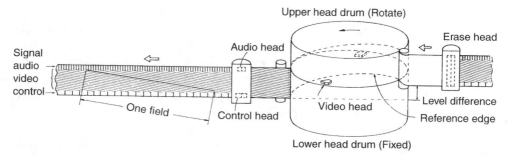

Signal
audio
video
control

One field

Upper head drum (Rotate)

Audio head

Control head

Video head

Lower head drum (Fixed)

Erase head

Level difference

Reference edge

Figure 13–4 Schematic format of a two-head helican-scan video tape recorder and tape system. [*Courtesy of Matsushita Electric Industrial Company.*]

mounted in tandem, one above the other, in the container, the tape between the two coaxial reels automatically has a slant position that can compensate for the helical wrapping difference. In the beginning of 1970, recording technology was still immature, and the tape consumption was too large to fit into one container, especially with a parallel reel configuration like a compact cassette (an hour of recording required a 7-inch reel).

Philips introduced a tandem cassette color video recorder (N 1500) to the European market in 1970. The machine could record up to one hour and had a timer clock that could set the time to record. The American company Cartrivision also used a tandem cassette in 1971 (Fig. 13–5). Using the field-skip method, which recorded only one field out of three, the machine could record up to

Figure 13–5 A tandem cassette recorder by Cartrivision.

3 hours, an epoch-making recording time for the early 1970s. Cartrivision marketed mainly prerecorded tapes copied from movies. A movie has 24 frames per second, and the field-skip machine could record 20 fields per second. Hence, the picture quality of the prerecorded movie tape was just acceptable. The prerecorded tapes were produced by a very new contact printing method developed by Panasonic in 1968. Cartrivision's field-skip method was not, however, suitable for TV programs and video camera recording, so marketing was restricted to a primitive form of the rental video business we have today. The machine was large and heavy, in keeping with the immature recording technologies of that time. In addition, after the tape had been used for some time, its surface became susceptible to chemical damage. Thus, this very interesting video system failed.

The last tandem cassette was developed by Matsushita Kotobuki (a subsidiary of Panasonic), which started to sell the VX-2000 video cassette recorder in a limited area of Japan. This video cassette recorder was a single-head, α-wrap using a 0.5-inch tandem cassette that had a space for the rotary head. One of the advantages of this video cassette recorder was easy production, provided programs were to be recorded and played back on the same machine. A single head required only a simple mechanism, and there were no difficult head positioning or pairing problems. But the cassette, with its two reels and a space for the rotary head, was bulky and rather expensive compared to the Betamax cassette, which was soon marketed in the same area. The parent company Panasonic decided to adopt the VHS system in 1976, and the VX-2000 was obsolete.

The style of cassette for a table-model video recorder was established as a parallel configuration by Betamax and VHS. It was similar to a paperback book in both size and appearance. The cassettes could be stored on a bookshelf but were too large to put into a portable or handheld video cassette recorder, and the

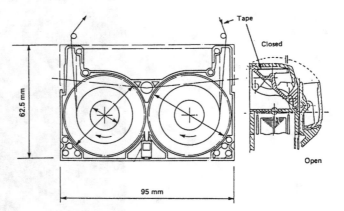

Figure 13–6   Schematic view of the construction of an 8 mm video cassette lid. [*Courtesy of Matsushita Electric Industrial Company.*]

tape was still partly exposed. The "8 mm video system" was established to work in this situation.

The 8 mm video cassette had the ultimate video tape cassette construction (Fig. 13–6). This cassette protected the very delicate and fragile tape from dust, fingerprints, and atmosphere by means of a very skillfully designed lid. Thus, the long history of the development of video cassettes was completed. Many other video cassettes developed later, especially for broadcast use, had essentially the same construction as 8 mm video cassettes.

BETAMAX (BETA) VERSUS VHS IN JAPAN

Many video cassette recorders were developed in Europe, the United States, and Japan before the oil crisis in 1973. Engineers applied their skills and efforts to design machines and appliances for low power consumption. In Japan at that time, the race to develop home video cassette recorders was just coming to the last stage. In 1974 Sanyo/Toshiba developed a 0.5-inch tape, 0.5–1 hour video cassette recorder (V-Cord). In 1975 Sony marketed the 0.5-inch, 1-hour Beta (Fig. 13–7), and Matsushita Kotobuki unveiled the VX-2000 in a limited geographical area.

In 1974 Sony tried to persuade Panasonic and JVC to adopt the Beta format. Unfortunately, the timing was poor. Each company was determined to develop its own video cassette format at that time. If Sony had tried to negotiate with Panasonic to develop a cassette jointly, the result might have been different. But Sony felt that joint development would take longer than an in-house project. Another important point was the recording time, in that Sony designed the Beta cassette size to accommodate one hour's recording. Panasonic, however, had learned from a market survey on video recorders that customers would want to record movies and sports programs, which last at least two hours. Japan Victor, too, had carried out market research that led to a target design of 2-hour recording in a small cassette. Eventually, JVC approached its parent company, Panasonic, to pursue this

Figure 13–7 The first azimuth color video cassette recorder, the Sony Betamax. [*Courtesy of Sony Corporation.*]

effort jointly. The result was what came to be known as the video home system, or VHS, format. The 2-hour recording format became the key issue, and once JVC had achieved a working prototype, the company was able, after negotiations with five major companies in Japan, to disclose the VHS format video cassette recorder, and market such a device in 1976 (Fig. 13–8). The VHS video cassette recorder was smaller than the Beta, in spite of having a slightly larger tape cassette. The secret was its tape-threading mechanism, which used simple M loading with two small pin guides.

Toshiba, Sanyo, NEC, and others adopted Sony's Beta, while Panasonic, Hitachi, Sharp, Mitsubishi, and Akai supported VHS. The Japanese market was just divided in half. This was the first attempt to create a big standard in Japan. Previously, major standards such as color TV systems and audio compact cassettes had been imported from the United States and Europe. The market accepted both Beta and VHS formats until the rental video business became popular, even though customers would have preferred a single format. The government tried to negotiate to unify the two formats, but this was impossible.

The Japanese market was the most severe test for cassette recorders. There were many manufacturers, and almost the same numbers of Beta and VHS machines were put on the market at the beginning. Sony is a very aggressive company, constantly striving to improve its machines. One of the biggest problems of both formats was audio sound quality. After the recording time increased by a factor of 3, the tape speed was only 1 cm/s (even the audio compact cassette runs at 4.75 cm/s). To improve the sound quality, Sony tried to record stereo audio by FM using the video head. To provide room for the audio FM carrier frequencies, the FM carrier of the video signal was shifted to a position 400 kHz higher. This was believed to be the maximum shift possible while still maintaining interchangeabil-

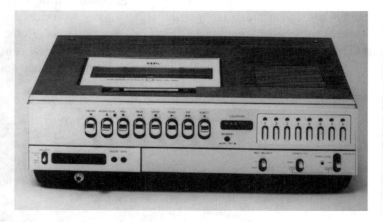

Figure 13–8   The first VHS video cassette recorder, the JVC HS-3300. [*Courtesy of Matsushita Electric Industrial Company.*]

ity with conventional machines. This carrier shifting served to eliminate a portion of the modulated video signals affecting the resolution of the picture, as well.

Panasonic invented a new FM audio recording system using two separate audio heads having gaps inclined at a large angle on the rotating drum. The audio head records the FM audio signal deeper into the tape at a 30-degree azimuth angle before the video signal is recorded on the same track. Because of the differences in azimuth angle, depth of penetration, and wavelength, the audio heads pick up only the FM audio signal, and the video heads only the video signal. This audio recording method left the original VHS tape format unchanged, which is very important for the prerecorded tape business. Soon afterward, users of FM audio Beta video cassette recorders complained that the insertion of the FM audio signal spoiled the picture resolution. Sony improved the picture quality by shifting the video FM carrier again to produce the high-band Beta. With this second improvement, Beta lost its interchangeability between old and new models. Thus, the Beta format destroyed itself. In 1988 Sony decided to adopt the VHS format.

BETA VERSUS VHS IN THE UNITED STATES

Sony had supported the sale of the one-hour Beta from Zenith in the United States in 1976. Panasonic tried to strike back in the U.S. market with the 2-hour VHS in 1977, dealing first with RCA, one of the biggest American consumer electronics companies. Sony then developed a means of switching between 1- and 2-hour recording time, and cut the track width in half. RCA pointed out to Panasonic that the tape cost per hour of the new Beta was now less than that of VHS. For 2-hour recording, the new Beta needed only 144 meters of tape, while VHS needed 237 meters. During the negotiations, the parties considered the programs most likely to be recorded on video cassette recorders. They agreed that, similar to the Japanese experience, the programs most likely to be recorded would be movies and sports. The requirements for recording sporting events differ in the two countries, however, since American football, one of the most popular sports, requires at least 3 hours. Panasonic proposed a 2- and 4-hour switchable machine, following the approach taken in the Beta. After several discussions, RCA decided to adopt a 2/4-hour switchable VHS system from Panasonic, having a retail price of less than $1000.

After RCA's decision, many other electronics companies such as GE, Magnavox, Bell & Howell, and Sylvania also adopted the 2/4-hour VHS. The Beta group companies had their own U.S. sales channels, but the VHS format gradually gained majority acceptance, and after the rental video business became popular, its market share accelerated.

BETA VERSUS VHS IN EUROPE

Beta and VHS both appeared in Europe in 1977. The European market for color TV was rather complicated because of the two color systems in use. As discussed in Chapter 9, most European countries had adopted the PAL system, the major exception being France. France improved the PAL system, to develop SECAM. Television receivers in Europe could be sold only under the control of the PAL or SECAM patents. Japanese TV set manufacturers were not able to sell their TV sets in Europe. At that time, video cassette recorders were just recording machines for TV programs. Japanese video cassette recorder manufacturers had no sales channels in Europe, and they had to sell their machines on an OEM (original equipment manufacturer) basis.

Beta and VHS stood on the same starting line in Europe. Fortunately, the VHS format, normally providing 2-hours of NTSC, the U.S. standard, could be modified to give 3 hours of PAL/SECAM with excellent color signal alignment and slightly longer tape length. Using the same method, Beta could easily modify its format to provide 1.5 hours of PAL/SECAM. To make a 3-hour PAL/SECAM Beta, however, the engineers had to change the position of the two heads from a simple 180 degrees to 180 degrees, 17 minutes, 17 seconds, to achieve alignment of the synchronous signal. Even so, the color signal did not line up straight. Although customers had no problems with this machine, it tended not to gain the approval of the OEM engineers. Thus, almost all TV manufacturers and sales companies in Europe decided to adopt the VHS system rather than Beta.

Philips did not accept the VHS format easily. The Dutch company developed its own machine, the V-2000 (Fig. 13–9), having a 0.5-inch tape cassette, almost the same size as VHS. The big difference was that it recorded on half of the

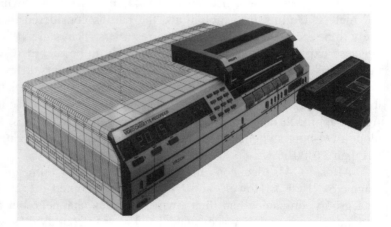

Figure 13–9   The Philips video cassette recorder, Model V-2000. [*Courtesy of Philips.*]

tape surface in one direction and on the other half in the reverse direction when the cassette was turned, an arrangement comparable to an audio compact cassette. The tape format was almost the same as VHS PAL/SECAM, but the track width was about half of that of VHS. Philips's claim that the V-2000 could record up to 8 hours on thinner tape sounded very attractive for customers, who want to record TV programs for as long as possible. However, the prerecorded tape suppliers hesitated to use this system. Each recorded program called for a different tape length, and it was necessary to turn the V-2000 cassette just halfway through the program, while the VHS system could record straight on with only adjustment of the tape length to the program recording time. Even Philips, which had developed so many successful home systems such as the compact cassette, laser disc, and compact disc could not predict the future of the rental video business. In 1983, Philips finally decided to adopt the VHS system. The European market was now united on one video cassette recorder tape format, VHS.

THE FIRST 8 MM VIDEO FORMAT CONFERENCE

Beta and especially VHS were very successful as tabletop recorders. Many customers bought them to record TV programs using timers. Sometimes, they wished to record programs from different TV stations, and the timer was soon improved to record seven or more programs from various stations at different times. Eventually, in 1980, the recording time of the recorders was increased to give up to 6 hours on a 19 μm track width, which is one-fifth the thickness of a dollar bill. When the International Electrotechnical Commission standard for VHS was established, the format had been proposed in only a 2-hour standard mode; 4- and 6-hour modes were just additions. Prerecorded tapes used only the standard 2-hour mode, which was adequately long and had enough tolerance to maintain machine-to-machine interchangeability. This is another reason for the great success of the VHS rental video business.

Another very important function of video cassette recorders was to record picture and sound through a video camera with microphone, and provide immediate playback. Filming with 8 mm movie film had neither sound recording nor instant playback. Every video cassette recorder manufacturer tried to develop a small recorder and video camera using a vidicon tube. The first of these came on the market in 1982, and a small cassette recorder and camera were built in one body as a camera/recorder in 1985 (Fig. 13–10). Since this machine could use conventional video cassette tape, it was very easy for users to play back the tape on their tabletop machines, upon returning home from their video-taking excursions. The recording time was adequate, and tape cost was reasonable. The first of what would become known as "camcorders" was relatively large and definitely not handy; only a strong person could comfortably handle a machine of this size and weight.

Figure 13–10   An early VHS camcorder, Panasonic NV-M1. [*Courtesy of Matsushita Electric Industrial Company.*]

In principle, home movie recording time is generally short. An 8 mm movie was, for instance, only 5 minutes per reel or cartridge. By limiting the recording time to less than an hour, the cassette for a camcorder could be smaller than conventional video cassettes. In 1980 Sony developed a "video movie" system using a very small cassette using 8 mm wide tape having 20 minutes of recording time (Fig. 13–11). Hitachi also disclosed the Magcamera, in which a cassette using 0.25-inch-wide tape (similar in size to an audio

Figure 13–11   The Sony 8 mm "video movie"camcorder. [*Courtesy of Sony Corporation.*]

compact cassette) gives 2-hour recording. Soon after, Panasonic announced the Micro Video System, using a newly developed thin metal-evaporated (ME) tape of 7 mm width.

In the past, the standard was established by the competitiveness of the system: in other words, the de facto standard was the surviving one. This time, things were to be handled differently. Initially, engineers from three companies met to discuss a unified format for the camcorder; then JVC and Philips, as format holders, joined the discussion in 1981. The resulting white paper for a unified format was disclosed at the 8 mm Video Conference in 1982, attended by representatives from every world company wishing to adopt the new format. Over a hundred companies were represented. This was the world's first attempt to establish by conference a standard for videotape format. Five working groups explored different aspects by means of experiments. It was remarkable that so many people of different nationalities and so many companies, while competing like enemies in the market, could cooperate and become friends in addressing items of common interest. The official written language of the proceedings was English, but the official talking language in the working sessions was Japanese. Many Japanese engineers enjoyed participating in this international conference. It was the first, and maybe the last, chance to use Japanese in an international conference. Finally, in 1984, this conference agreed on the "8 mm video" format, the first international video cassette recorder standard formulated by conference.

THE VHS COMPACT CASSETTE (VHS-C)

At the time of the 8 mm Video Conference, the VHS format had become very popular as a tabletop model. The 8 mm video format began as a camcorder application, which would use a very small cassette. Nonetheless, the original specification was for a 2/4-hour recorder, the same options as for VHS. Manufacturers of VHS systems were worried about possible competition between VHS and 8 mm video. In this situation, JVC developed a compact cassette known as VHS-C. Comparable in size to a cigarette pack, the VHS-C could be inserted in an adapter, which could play it back as a conventional VHS cassette (Fig. 13–12). The recording time was 20 to 60 minutes. Both 8 mm video and VHS-C could record and play back audio/video signals with the same machine. The only remaining task was to provide a connection to the TV receiver. Television manufacturers quickly noted this point and put an audio/video input terminal on the front of their receivers. Now, the format was no longer a problem. Generally, however, customers came to prefer longer recording time and smaller size, and the 8 mm video format began to gain the majority share of the camcorder market.

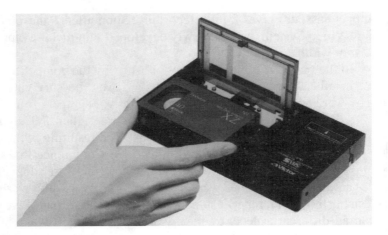

Figure 13–12   JVC's VHS compact cassette and adapter, the VHS-C. [*Courtesy of Matsushita Electric Industrial Company.*]

THE DEVELOPMENT OF VIDEO TAPE

Video tape corresponds to the film of a photographic camera. The grains of magnetic pigment in the tape should be as fine as possible, to increase the picture resolution. The magnetic strength (magnetization and coercivity) should be as high as possible, to increase the output at very short recorded wavelengths. Video tape is the most sophisticated of all magnetic tapes, and producing a high-quality product is difficult. The tape must contact the heads intimately via a very smooth surface, yet run with uniform velocity on a rotating-head drum.

The tape used for early fixed-head video recorders was just the existing audio tape. The main problem was to record very short wavelength signals. The only solution was to make the tape surface as smooth as possible. Development of tapes specifically designed for video started when the quadruplex recorder was introduced, then took off in new directions to satisfy the special needs of helical-scan recorders.

The use of very narrow track widths on early home video recorders made head life as well as short wavelength recording major problem areas. The tape surface had to be ultrasmooth to record short wavelengths, but a smooth surface hindered the uniform running of the tape on the rotating head drum because of an effect called "stiction." Until then, video tape was made using established coating technology to lay down a thin layer of fine magnetic pigment held in a binder. Coated tape had many advantages, such as easy production, precise control of tape runnability by lubricants in the binder, and a hard surface to protect against the atmosphere and resist abrasion by the head. Yet, users of magnetic recording technology wanted still higher recording densities.

In theory, higher coercivity and higher remanent magnetization were important for improved performance. Tape manufacturers had changed the magnetic pigment from ferric oxide to chromium dioxide to increase the coercivity. But chromium dioxide was expensive, and tape manufacturing companies, including TDK and Fuji Photo Film, developed an alternative, high-coercivity, magnetic pigment by the adsorption of cobalt onto ferric oxide. This tape was less expensive to produce than chromium dioxide and was just in time for the Beta and VHS era. The combination of cobalt-adsorbed, ferric oxide tape and ferrite head was very good. However, the manufacturers of the next generation of video cassette recorders were looking for another high-performance magnetic tape.

The strength of a signal recorded on tape is related to the magnetic properties of the magnetic ingredient of the tape. Magnetic metals such as iron or cobalt are more potent magnetically than any of the magnetic oxides and potentially could retain a higher level of recorded information. A Japanese company, Kanto Denka, developed a metallic iron pigment and made tape from it, calling it "metal tape" (MP). This metal tape was used first in audio cassettes and gradually applied to video. The first video application was to the Panasonic M-II format video cassette recorder described in Chapter 12.

Even with this "metal tape," the pigment was coated in a binder. A binder is very useful, not only for holding the fine particles of pigment together, but also for incorporating lubricants and materials for cleaning the video head surface. But the binder, being nonmagnetic, limits the strength of the magnetization obtainable. From a magnetic point of view, the ideal tape would be one in which the magnetic material is directly coated on the substrate without using a binder. To realize this kind of tape, experimenters tried to deposit cobalt directly onto the tape substrate in a vacuum chamber. Panasonic developed such a tape, known as "metal evaporated" (ME) tape, in 1978. Once again, the new tape was initially used in audio cassettes, but it was not very successful because it did not support the recording of the very long wavelength signals necessary for audio recording. As the saying goes, "a new wine needs new skin," so the tape was used for a completely new format called the Micro Video system.

Samples of ME tape were marketed first on the 8 mm video system by Kodak, supplied by Panasonic on an OEM basis. Unfortunately, the technology was not yet mature, and the tape tended to stain and curl. Panasonic withdrew from further development of ME tape for a long time, although the development was ultimately successful. Sony also developed ME tape and applied it to the 8 mm video High-Band series. This was the first commercially successful ME tape, and it is now used for digital video cassette recorders.

HOME DIGITAL VIDEO CASSETTE RECORDERS

Digital technologies are developing in step with the progress in semiconductor devices. Nowadays, many systems are changing from analog to digital. Examples are digital audio compact discs, digital audio tape, digital video tape, and digital TV systems (ATV and DVS). Home digital video cassette (DVC) recorders are designed to record not only analog camera/TV signals but also digital television signals directly on the tape. Initially, the main application for a home DVC was as a portable camcorder video cassette recorder, to replace the 8 mm video cassette recorder. The sensitivity of a camcorder camera is very high, and it has long since replaced the 8 mm home movie system. Even at night, one can take reasonable color pictures under low artificial light, by making use of advanced charge-coupled device (CCD) technology. What is more, the small size of CCDs make a zoom lens easy to package in a small volume. Thus, very high performance camcorders are in use all over the world. Despite their many attractive features, recordings made using camcorders are difficult to edit. If the desired editing is at all complicated, one scene must be duplicated several times or generations. An 8 mm video program has just enough performance to be enjoyable after it has undergone one or two generations of editing. Any further editing will seriously deteriorate the user's enjoyment of the recorded material. The reason is that 8 mm video or VHS-C camcorders are analog recorders, and their picture quality degrades almost three times (10 dB) after several duplications. The picture becomes very noisy and the resolution is also poor.

In principle, digital video systems have no degradation. At reasonable signal-to-noise ratios, the digits containing the picture information are reconstructed perfectly after each generation. Conventional digital recording requires a high sampling frequency, typically 13.5 MHz, with 8 bits of resolution. Although recording density is increasing every day, it is still difficult to record a one-hour program in a cassette smaller than audio size. In realizing this performance, the main contributions have been due to semiconductor technology, digital compression techniques, and advances in recording technology using new ME tape. Thus, a home DVC can record a 1-hour program in a small cassette and a 4-hour program in another big cassette. In 1996 the small cassette portable home DVC was put on the market. Immediately, home movie enthusiasts, as well as professional video production workers, started to use it. Recently, many freelance journalists have begun to use the DVC as an electronic news gathering device to record up-to-the-minute news from all over the world.

In the near future, tabletop machines will be put on the market for home DVC use, but many people will still select the VHS system because of the inexpensive tapes and abundance of prerecorded tapes. One of the exceptions will be the personal computer enthusiast. All information in the DVC can be transferred into a personal computer and used as a still or moving picture. Editing will be very easy, comparable to using a word processor. In the beginning of the next century, the standard home video recorder will be the DVC, provided personal computers are widely distributed and the cost of the DVC is suitably reduced. There are, however, many remaining problems to be solved, such as copyright protection, profitability, and the cost of prerecorded tapes for home DVC use.

TAPE DUPLICATION

The importance of prerecorded tape has been mentioned many times in this chapter. At the beginning of 1980, only very limited programs were copied on a small scale. To duplicate these programs, several video cassette recorders were recorded together from a master video cassette or tape recorder in real time. This technique was known as the head-to-head method. At the beginning, tape duplicators had to employ many technicians for system maintenance and quality control.

As an alternative, in 1969, Panasonic explored the possibility of using "contact printing" as a means of high-speed video duplication. This process functioned just like a photographic film printer. Instead of light, an ac bias magnetic field was applied to a master tape and a copy tape, which were in close physical contact. The master tape, having a higher coercivity, was not erased by the bias field, whereas the magnetization pattern on the master was perfectly recorded onto the copy tape by the linear (anhysteretic) property of the bias field. Because of the contact geometry, the pattern actually recorded was the mirror image of the master recording, but this was corrected in the making of the master.

Panasonic developed a contact duplicator system for the EIA-J cartridge in 1974. One of its applications was in certain "do it yourself" chain stores in the United States, where it was used to duplicate self-help instructional videos. This duplication system became obsolete in 1977 when these cartridges were replaced by VHS. Panasonic then produced a duplication system for VHS. Because this system was ahead of its time, however, it failed to achieve wide market acceptance. Ten years later it found a place as a means of reusing previously duplicated cassettes.

Sony also developed a contact duplication machine, Sprinter, in 1985. This system can duplicate from a master tape to a 2500-meter-long copy (a "pancake") by a contact duplication method running 90 times higher than normal speed. The master tape can be used more than 3000 times, but the picture quality is slightly degraded from that of the head-to-head method. This system is in use mainly for making inexpensive prerecorded tapes such as for children's programs in the United States.

Today the enormous volume of prerecorded videotapes is almost entirely duplicated using the original head-to-head method. This would have been thought unbelievable in the early days of video cassette recording but, after mass production of many video cassette recorders by many companies, the machines matured, and their reliability became very high. Head life was improved by several orders of magnitude. Some companies developed extradurable machines for high-quality duplication. Conventional quality control methods were shown to suffice.

Big duplication companies are now connecting a thousand specially designed VHS recorders to a master video tape recorder, which might be a Type C broadcast video tape recorder or a digital video cassette recorder. One program can produce a thousand copy tapes simultaneously, but only in real time; a 2-hour program requires 2 hours for duplication. If a million duplicated tapes are required, the operation is simply repeated a thousand times, even though copy tape insertion and removal is manual or semiautomatic at best.

The most advanced duplication method is double-speed pancake tape duplication. As mentioned before, it is very difficult to increase the head rotating speed even two times. Synchronous operation of the master and all the many copy machines together to maintain tape speed and head rotating speed is extremely difficult. However, Rank (USA) developed a double-speed duplication system in 1990. Videotape duplication has became a major business, operating in large factories and producing huge volumes of prerecorded cassettes.

REFERENCES

Houghton, W. D., et al., "A Simplified Video Tape System for Home Use," RCA International Division, License Operation, Engineering Report ER-39 (1962).

Sugaya, H., "The Video Tape Recorder: Its Evolution and the Present State of the Art of VTR Technology," *J. SMPTE*, **95**, 301–309 (1986).

Sugaya, H., "Video Recording," Chapter 5 in *Magnetic Recording Handbook*, C. D. Mee and E. D. Daniel, Eds., McGraw-Hill, New York, 1990.

Sugaya, H., "Analog Video Recording," Chapter 5 in *Magnetic Storage Handbook*, 2nd ed., C. D. Mee and E. D. Daniel, Eds., McGraw-Hill, New York, and IEEE Press, Piscataway, NJ, 1996.

Sugaya, H., et al., "Magnetic Tape Duplication by Contact Printing at Short Wavelength," IEEE Transactions on Magnetics, **MAG-5**, 437–441 (1969).

Sugaya, H., et al., "Standard Alignment Tape Recorder for EIA-I Type 1 Videotape Recorder," *J. SMPTE*, **83,** 901–904 (1974).

Sugita, R., et al., "Co-Cr Perpendicular Recording Medium by Vacuum Deposition," IEEE Transactions on Magnetics, **MAG-17**, 3172–3174 (1981).

14 Digital Video Recording

Koichi Sadashige

As earlier chapters have related, when commercial television broadcasting began in the United States in 1947, broadcasters faced the problem of how to accommodate the four time zones of the country. The solution was first provided by the quadruplex video tape recorder, developed by Ampex Corporation in 1956. Later, the quad machines were replaced by more economical field-per-scan, helical-scan recorders, such as the Type C equipment in use today.

Analog video recorders exhibit several shortcomings, notably a steady increase in the random noise level, color differential gain and phase error propagation, gradual loss of the luminance signal bandwidth, and lack of multiple audio channels. These deficiencies have become apparent to the more critical video recorder users, such as the producers of high-quality commercial advertisements. The teleproduction industry began to demand a better solution, looking beyond the Type C recorder capability and performance.

DIGITAL RECORDING OF ANALOG SIGNALS

The First Proposal for Digitization of an Analog Signal

As early as 1937, A. H. Reeves of England, while working at the Paris Laboratories of the International Telephone and Telegraph Corporation, proposed the use of pulse code modulation (PCM) for bandwidth conservative transmission of multiplexed telephone channels. Reeves argued that PCM is a better way to transmit a multitude of telephone voice channels over a single transmission medium than frequency multiplexing, the technique then used by the industry. The world, however, had to wait until after the second world war and the invention of the transistor for Reeve's proposal to be reduced to practice.

In the 1950s, AT&T Laboratories began development activities on multiplexed telephone transmission equipment based on PCM technology. For the first time, an analog signal was purposely being transformed into a digital data stream. The obvious benefit to the electronics world outside the telephone industry consisted of the accelerated developmental efforts on analog-to-digital converters, and their eventual commercial availability. The capability of the first-generation analog-to-digital converter (ADC), useful only for a narrow-bandwidth voice signal, had advanced by the late 1950s to the level of operating with sampling frequencies in hundreds of kilohertz.

Attempts to Digitize Sound Recording

The music recording industry has always been influenced by purists who can detect the most minute degradation of sound quality. For a prolonged period after 1950, full-track recording on 0.25-inch magnetic tape at 30 in./s, a product of the postwar audio equipment industry, was judged to be adequate for original master recording and subsequent editing operations.

The music industry, never satisfied with the status quo, had been looking for a recording system with a much greater dynamic range and wider bandwidth than the best analog system can offer. The stated requirements were a 90 dB dynamic range and a bandwidth in excess of 20 kHz.

Also demanded was complete processing transparency as the original master recordings are edited through multiple generations to produce the final release copies. The only solution to the industry's demand seemed to be the use of as yet unexplored digital recording. Experimentation on digital audio recording, then referred to as PCM recording, dates back to the early 1960s.

For digital audio recording, with limited data rate requirements, the recorders then being produced for computer data storage and instrumentation recording were technically more than adequate. By the mid-1960s, several laboratories and magnetic tape recorder manufacturers were demonstrating experimental digital audio recorders, in which high data rate recording technology and PCM processing of the audio signal were combined.

The lower cost, compact video tape recorder known as the U-Matic system, came into the marketplace in the late 1960s for electronic news gathering (ENG) and other professional applications. Converted to digital audio recording, this equipment also provided exceptionally good reproduction quality.

Digital audio recorders for professional applications, based on stationary heads and multitrack configuration, entered the marketplace in the 1970s. Some of these recorders used the first-generation multitrack magnetoresistive heads for playback operations. These early activities laid the foundation for the later development of the CD (compact audio disc) and DAT (digital audio tape) products.

Digital Conversion of a Video Signal

The NTSC and PAL color television systems were described in Chapter 9. Analog recording of PAL composite color television signals using a frequency-modulated carrier produces a higher level of unwanted aliasing components because of the higher color subcarrier frequency of 4.43 MHz. This results in a poorer video quality compared with NTSC composite signal recording. It is therefore understandable that in the early 1970s, two major video recorder users of PAL, the Independent Broadcasting Authority and the British Broadcasting Corporation, both of the United Kingdom, began to investigate the feasibility of digital video recording. Video recorder manufacturers in the United States, Europe, and Japan, under mounting pressure from the teleproduction and broadcasting industries, had also started feasibility studies.

The digital conversion of a video signal is decidedly more complex than the conversion of audio signals to the digital domain, as described earlier in this book. In digital audio recording, a sampling frequency of some 48 kHz, and quantization of 16 bits per sampling can be used, giving a data rate of 768 kb/s. For a digitized video bit stream carrying a composite color video signal, data rates in the 100 Mb/s region or higher are required.

The successful execution of digital video recording required the development of two technologies: analog-to-digital and digital-to-analog conversions fast enough to cope with the bandwidth of the broadcast-quality video signal, and the short wavelength/narrow track recording needed to provide an acceptable rate of recording media consumption. Tracing the origin of these technologies reveals some interesting facts about the growth and development of electronics in the period before, during, and after the second world war.

When the PCM technique, originally proposed in 1937, was pursued by the U.S. Army in the early 1950s, one intended application was the encryption of telephone communications. Analog voice encoding techniques, developed in the 1930s and 1940s, were never completely secure despite a high degree of complexity. Digital coding, with its coding algorithm arbitrarily alterable via software modifications, was the only secure voice encryption method. The digital conversion of a video signal has a similar origin, its first application probably being for U.S. national defense purposes.

In the early 1960s, the U.S. government had a color television network connecting major governmental installations across the North American continent. Coaxial cables connecting these installations were not entirely tap-free, and there was always a possibility of transmission interception by unfriendly parties. As is the case with telephone channels, the analog coding of video signals is insecure. The only viable means of video signal encryption is the use of digital technology. Thus, efforts were under way as early as the mid-1960s to develop high-speed, video-rated, analog-to-digital and digital-to-analog converters.

Coincidentally, needs for the video-rated ADCs and DACs were also developing in the commercial sectors.

Digital Islands in the Analog Sea

A video tape recorder is a complex electromechanical device, and its output signal contains timebase instability. The timebase instability is generated through the frequency and phase nonuniformity of the head scanner disk, variations in the tape speed, and the elasticity of the recording medium. If the output is to be used for program editing and mixing, and if it is to be switched into programs from other sources for television station on-air operation, the video recorder output signal must be perfectly synchronized to an outside timing reference.

Earlier designs of timebase correctors, which reduced the output signal timebase instability and synchronized it to an outside timing reference, used analog delay lines. Because the timing compensation range available in the analog delay line was restricted, the timebase corrector of the 1960s and 1970s was of limited utility for many applications. With the advent of semiconductor digital memory, especially dynamic RAM, with its increasing data storage capacity, it became obvious that the next generation of the timebase correctors should be based on the digital memory. Since the need for extended-range timebase correctors was growing rapidly in the early 1970s, experimentation with high-speed ADCs for broadcast and teleproduction applications began in earnest.

Digital technology offered another important tool to the teleproduction industry: "digital special effects." Once an image has been transformed into a matrix of pixels, it is possible to alter its size, aspect ratio, brightness, color, and linearity. All these alterations can be done on the total image, or on a portion of the image. The field was then wide open for digital manipulation of video images. Ampex Corporation, one of the the developers of the original concept, called the technology *Ampex Digital Optics*. The teleproduction industry uses the term "digital special effects," or "digital video effects." Growth of digital video equipment applications such as the digital timebase corrector, and the digital special effects equipment within the otherwise entirely analog environment of teleproduction and broadcasting plants was referred to as the "digital islands in the analog sea."

In August 1974 the Society of Motion Picture and Television Engineers (SMPTE) established a study group on digital television, which carried out valuable work including the selection of acceptable signal sampling rates and the quantization of word sizes.

High-Density Recording Technology

A quadruplex recorder has a channel bandwidth of approximately 15 MHz in the high-band operating mode, which can be equated to a digital data transfer rate of approximately 30 Mb/s. At the recording rate of 100 Mb/s, the required digitized composite color video signal data rate estimated by earlier studies, the digital recorder would be using the tape at a rate three time as fast as the analog recorder, and obviously this would be entirely unacceptable. It was necessary to find a means of greatly increasing the recording density. The digital video recorder developers found sources for the needed high-density recording technology, not in the computer industry, but in instrumentation recording and in the low-cost, video helical-scan recorders just entering the marketplace.

Instrumentation Recording

The end of the war in 1945 did not bring the expected pause in the development of U.S. military aircraft. In fact, postwar activities encompassed not only aircraft but new warhead delivery systems and guided missile development, design, and production. To monitor and record the in-flight data of the new aircraft and missiles, high-performance data acquisition and storage systems were urgently needed. By advancing the then emerging audio recording technology to its practical limit, engineers succeeded in designing the first generation of instrumentation recorders as single or multichannel stationary-head equipment recording both analog signals and digital data streams. Instrumentation recorders are inherently high recording density devices. Low media usage, compact size, and light weight were their important features.

Later, quadruplex video recorders, with extensive mechanical and electrical modifications to operate at a data recording rate as fast as four times the standard video version, were produced to meet the newly developed requirements of electronic intelligence and remote sensor output data recording.

These and other similar instrumentation recorders offered highly capable platforms for initial digital video recording experimentation.

Running on a Mile-Long Bridge
That Is Only 2 Feet Wide

Lower cost video tape recorders and some high-density instrumentation recorders were based on the helical-scan technology. The tape, wrapped around a stationary or rotating drum a few inches in diameter, moves at a slow linear speed. Heads are mounted on a disk or a drum of the same diameter as the stationary drum.

The head scanner rotates at a high speed, thus giving the recording system the high head-to-tape velocity needed for a high bandwidth or a high data transfer rate.

The length of the head track, or the helical track, is normally several inches. To minimize tape usage, the width of the track is made as narrow as practical. If the track is 5 inches long, and the head is 0.002 inch wide, the ratio of the track length to the width is 2500:1. The head is moving along the track at a linear velocity in the region of 300 to 1000 in./s. The situation is analogous to driving on a mile-long bridge, which is only 2 feet wide, at 100 mph.

The technology developed during 1970s to keep the head on the long and narrow track is the basis of the successful development of the digital video tape recorders of later years.

The head tracking technology is in constant development to meet the quest of the marketplace for ever higher recording density. The technology involves both the refinement of the mechanical accuracy of the entire head-scanning system, as well as employment of an automatic head placement system.

THE ROAD TO D-1

Initial Experimentation on Digital Video Recording

By the early 1970s, the basic technologies needed for digital video recorder development, video-rated ADCs and DACs, and high-density recording, had become available, albeit in somewhat elementary form. One of the first proposals for a digital video recording system came from the Independent Broadcasting Authority (IBA) of the United Kingdom in the form of a technical paper given at the Conference on Video and Data Recording at the University of Birmingham in July 1973. John Baldwin had argued that a digital video recorder with tape usage equal to the quadruplex system (i.e., 30 in.^{2}/s) could be designed to record a PAL composite color video signal. The signal would be sampled at three times the color subcarrier frequency with 8 bits per sample, at the user data rate of 106.4 Mb/s. IBA carried the development to completion and demonstrated a working prototype based on the Model 9000 broadcast helical-scan recorder of International Video Corporation. IBA also developed a channel code concept based on the use of selected longer length words (e.g., 10-bit words) to replace the original shorter length 8-bit words, for a gain in recording density. This was the forerunner of 8-10, 8-12, 8-14, and other similar channel codes.

By the mid-1970s, laboratories throughout the world were in a frenzy of activity experimenting on digital video recording, with both rotary-head (quadruplex and helical-scan) and stationary-head/multitrack recorders. Ampex built the initial digital recording technology using its quadruplex expertise and, by the late 1970s, was in a position to demonstrate a fully operational digital recorder.

In digital recording, the video signal need not to be recorded as a single continuous bit stream. Therefore, a multitrack digital recording is a definite possibility. The BBC had taken this approach and, in 1975, built an experimental stationary-head, multitrack recorder for a sub-Nyquist sampled PAL composite signal (Fig. 14–1). The recorder laid down 42 parallel tracks on 1-inch wide tape at an in-track density of 15 kb/in. The user data rate was 75.6 Mb/s.

In 1976, NHK (Japan Broadcast Corporation) technical research laboratories built an interesting exploration tool for digital recording technology using a rotating magnetic sheet (Fig. 14–2). Rather than building a product prototype, NHK wanted to accumulate basic technical information on high-density recording, error correction techniques, and head–medium interface conditions.

The NTSC composite signal was sampled at three times the color subcarrier frequency, with 8 bits per word, forming a bit stream at 86 Mb/s. The bit stream was divided into two channels. With the addition of error correction overhead, each channel operated at 48 Mb/s. Two channels were simultaneously recorded on the rotating sheet at a linear track density of 16.9 kb/in.

The mid-1970s was a time of aggressive activities on broadcast-quality, helical-scan video recorder design and manufacturing. Bosch–Fernseh of Germany, Sony Corporation of Japan, and Ampex and RCA of the United States

Figure 14–1 Longitudinal multitrack magnetic tape digital video recorder. [*Courtesy of the BBC Research Department.*]

Figure14–2   Rotating magnetic sheet digital video recorder. [*Courtesy of NHK Technical Research Laboratories.*]

had used helical-scan recorders of various forms for their respective experiments on digital video recording. The range of recording density used in the experimental digital video recorders was considerably higher than the high-band quadruplex system, representing significant advances in the recording medium, the record/reproduce head performance, and the signal electronics. The track pitch ranged from 0.004 to 0.006 inch (100–150 μm) and the in-track density was between 15 and 20 kb/in. (1.7–1.2 μm/b). The high-band quad system had a track pitch of 0.015 inch (170 μm) and the equivalent in-track density of 12 kb/in. (2 μm/b).

Digital Video Recorder Standardization Activities

Digital recording formats proliferated in the late 1970s. By 1979, both user groups and potential recorder manufacturers realized that they would benefit enormously from the establishment of an industry standard for a digital video recorder that represented the state of the technical art while meeting the requirements of the majority of the users.

In the United States, the SMPTE had formed a new study group with the expressed purpose of studying the technical feasibility of digital video recording. In

Europe, the technical committee of the European Broadcasting Union (EBU) had established a specialist group with the same objective from the users' view point. The EBU group was called MAGNUM (**mag**netoscope **num**érique). The next three years, which constituted the formative period for digital video recording technology, were marked by countless separate and joint SMPTE and MAGNUM meetings, attended by experts from the United States, Europe, and Japan. The digital video recorder configuration, which was being developed, was intended primarily for television program production. The unified 525/60 and 625/50 component digital video standard (referred to as Recommendation 601), also developed by SMPTE and endorsed by its European counterpart, the Comité Consultatif International des Radio-Communications (CCIR), was the selected input/output signal format for the recorder.

By 1983, the work of the study groups indicated that development of the actual recording format, along with the recorder and recording medium specifications, appeared feasible. Early in 1984, SMPTE reorganized the study group into the Working Group on Digital Television Tape Recording (WG-DTTR), with the specific assignment of developing the recording format, together with tape and cassette specifications. SMPTE draft documents for the DTTR system, the D-1 format, were completed by late 1985. Later, the format was accepted by CCIR and the International Electrotechnical Commission (IEC). The D-1 format established three cassette sizes, with an increase of approximately 2:1 in the data capacity between sizes. The tape is 19 mm (0.75 in.) wide, with a high-density oxide coating.

The First Digital Video Recorder

By late 1986, two manufacturers, Bosch–Fernseh of Germany and Sony of Japan, were offering first-generation D-1 recorders. The recorders established a rare example of a truly successful product despite its origin in committee activities. Because the D-1 recorder was the first volume-produced, high-data-rate digital recorder, its usefulness in applications outside the television industry was well appreciated even before the format specifications were finalized. User groups in U.S. government circles, the instrumentation recording industry, and NATO organizations were among the first to see the obvious cost advantages of the volume-produced, industry standard product over the prevailing practice of purchasing small quantities of specially designed recorders. The instrumentation recording format ID-1, U.S./NATO military recorder specification MIL-2197, and the data recorder format called DD-1 are all derivatives of the basic D-1, a testimony to the high level of technical perfection of the committee product.

Although new demands for the D-1 video recorder have tapered off in recent years, ID-l and DD-1 format equipment is still in series production in Japan by Sony and in France by Schlumberger.

COMPOSITE VIDEO DIGITAL RECORDING

The Need for a Composite Video Digital Recorder

The D-1 recorder, being the first-generation product of its kind, and also operating at a high data rate to accommodate the component video signal, was expensive to purchase and operate. The broadcast industry's need for digital recorders was growing rapidly in the early 1980s. The industry, however, demanded a recorder that could be acquired and operated at lower cost.

The concept of composite digital video recording then emerged among equipment manufacturers, and U.S. and Japanese broadcasters. Around this time the quadruplex recorder, the old standby of broadcast recording, was being replaced by the Type C and Type B (mostly in Germany) 1-inch, helical-scan recorders. Although the 1-inch, helical recorders delivered the expected improvements in overall video quality, they still lacked image quality transparency beyond about the fifth generation. Also, a trend was developing among broadcasters to produce programs, commercial messages, and promotion materials in-house, rather than contracting out to teleproduction firms. For these purposes, the use of digital recorders was preferred.

The digital recorder configuration developed by Ampex for the broadcast industry records the composite color video signal, sampled at four times the color subcarrier frequency, with 8 bits for each sampled pixel. Ampex also elected to use the newly developed metal particle tape as the recording medium. Thus the company achieved a higher areal density, using a narrower track pitch and a shorter wavelength than the D-1 system. Unlike D-1, the new recorder also used the azimuthal head gap concept, eliminating intertrack guard bands.

Sony Corporation had joined the Ampex efforts in mid-course. The outcome of the joint company efforts was the SMPTE D-2 format, which was adapted by both SMPTE and the International Electrotechnical Commission (IEC) in 1988–1989. The D-2 recorders adopted the D-1 cassette shell design and quickly became the mainstay of high-quality broadcast recording throughout the world. These recorders remain the standard means of program exchanges among all Japanese commercial television stations. As in the case of D-1, the advantages of the D-2 recording system were immediately recognized by the instrumentation and data recording industry.

Today, the data storage versions of D-2 recorders are used for the storage of important U.S. government-generated geological and geospatial information, as well as other data storage applications.

The Quest for a Single Recording Format for All Broadcast Applications

Recording equipment is perhaps the most important and most widely used hardware in a television plant. Recorders are used for program acquisition, production in the field and in the studio, postproduction, archival storage, and on-air operations. In the past, these varying applications were met with a variety of recording equipment, from 2-inch quadruplex to consumer VCRs. The existence of a myriad of different recording systems and formats is an operational burden a television plant cannot support for long. NHK was at one time operating nearly 3000 items of video recording equipment of various types. Replacing all this recording apparatus with equipment conforming to a single, interchangeable digital format was the ambitious objective NHK set in the early 1990s.

The 0.75-inch tape standards of the D-1 and D-2 formats were unsuited to meet one essential requirement of NHK: namely, to have a shoulder-mounted camera/recorder combination ('camcorder'), of reasonable size and weight. A higher recording density technology, based on 0.5-inch-wide tape, was developed to meet the requirements of a single format for all applications. The technology developed by NHK was transformed in 1993 to a line of products made by Matsushita and, in due course, the new 0.5-inch format became yet another digital format to bear a SMPTE designation, the D-3. The D-3 product family, in production today, includes a camcorder, general-purpose tabletop models, and library systems of varying capacities.

The D-3 video recording format was also adapted by the data recording industry as the SD-3 format, with format modifications mainly in the area of error correction code structures.

DATA COMPRESSION

Emerging Needs for Data Compression

In the 1980s there were three competing and incompatible consumer video recording formats: VHS, Betamax, and 8 mm. Proponents of the VHS and Beta formats had begun looking beyond analog recording and had started experimentation on low data rate, but very high density digital recording technology as the basis for the next generation of consumer recorders. By the late 1980s, with the advances in display technology, the difference in quality between the best received video and standard video cassette playback had become noticeable. Demands for a higher quality cassette recorder were expected to develop in the consumer marketplace.

Compression of the original digital bit stream to a manageable data rate of 20 to 30 Mb/s was thought to be necessary to keep the tape usage rate and the cost acceptable for consumers. There was also another driving force for the utilization of data compression technology among the recorder manufacturers. This was the expanding broadcast electronic news gathering (ENG) equipment market. In the 1980s, all such equipment was analog, but the need for higher quality recording was developing. Local news programs were often the main means of identifying a particular station in the market, and a station rating was heavily influenced by the degree of acceptance enjoyed by its local news program. For ENG applications, the recorder must be light and small and, to cover the broader market segments, the recorder should be even smaller than the D-3.

This time period was prior to the advent of widespread video data compression activities represented by the formation of JPEG, the Joint Photographic Experts Group, and MPEG, the Moving Picture Experts Group. Recording equipment manufacturers, therefore, worked independently to develop their own compression technology. The companies involved included Ampex for ENG applications, and Matsushita, Philips, and Sony for the consumer products.

The technique explored extensively during this period was differential pulse code modulation (DPCM), where only the differences between the preceding and present samples were recorded. The technique is quite relevant because of the generally gradual rate of image content shifting from one sample to the next, and the high degree of information redundancy in the images. A refinement of the basic DPCM is to extend the concept further and operate the process not on a pixel-by-pixel basis but on a block containing 16 or more pixels. The Hadamard transformation, which yields a compression ratio of approximately 2, was also explored and used in experimental recorders.

In 1987 Philips had pioneered the use of discrete cosine transform (DCT) as a means for video data compression in an experimental consumer cassette recorder. No commercial products, however, emerged from these earlier recorder developmental activities using data compression. It was not until 1992 that consumer video cassette recorders with data compression began to be developed on an industry-wide basis.

Data Compression in Broadcast and Teleproduction Recorders

The original application for video data compression technology was to provide a means of delivering high-quality video material to the end user through a bandwidth-restricted transmission channel. The technology was not intended to be used during program production operations, which include recording. However, the advent of the MPEG and other compression techniques attracted the interest of video recorder designers as a way to reduce the media usage rate, the major recorder operating cost. For broadcast and teleproduction applications, preservation of the original image quality is a paramount user requirement. Therefore, although a high compression ratio will yield large media cost reduction, the ratio must not to be so high that it compromises the output image quality.

The first high-grade professional video recorder to use data compression was the Ampex DCT (digital component technology) series of recorders introduced in 1992. Derived from the D-2 mechanical configuration, Ampex used a data compression ratio of approximately 2. The recorder tape usage rate, with an input component digital video data rate of 216 Mb/s, is the same as the D-2 recorder with its input composite digital video data rate of 115 Mb/s.

In 1993 Sony entered the market with a digital successor to Betacam, the Digital Betacam. This digital component video recorder uses a data compression ratio of 2 to keep tape usage rate low. The Digital Betacam uses tape with superior magnetic properties, running at 96.7 mm/s—nearly 20% lower than its analog counterpart.

The latest addition to the Betacam series of ENG recorders, the Betacam-SX, was introduced in 1996. With a relatively high compression ratio of 10, it records the video data stream at 18 Mb/s, together with four uncompressed digital audio channels. It uses metal particle tape running at a speed only half that of its analog counterpart.

The determination of a compression ratio acceptable to demanding applications such as high-quality teleproductions has entailed much discussion. Because of the stringent industry demands for frame-by-frame editing, only the data compression algorithms operating within the television field or frame boundary are acceptable. This requirement places severe restrictions on the degree of compression achievable without creating discernible image degradation. By the mid-1990s, however, advanced compression algorithms made it practical to compress the video data by a ratio as high as 4 or 5, with acceptable image transparency through several coding/decoding, or compression/expansion sequences.

HDTV DIGITAL RECORDING

Digital Recording of a High-Definition Television Signal

High-definition television, HDTV, broadly refers to a television scanning format in which the resolution in both the horizontal and the vertical directions is approximately twice the current standard. In addition to the higher spatial resolution capability, an HDTV image is generally presented in a wider aspect ratio—a wide-screen picture—and it is usually accompanied with 3-D-type sound effects, often accommodating more than one language.

NHK is the first broadcasting organization in the world to broadcast HDTV on a regular basis. The NHK HDTV system, referred to as "Hi-Vision" in Japan, requires a data rate of 1.188 Gb/s to carry its high-resolution wide-screen image. In the late 1980s, NHK had undertaken the task of developing a digital recording system, using Type C tape transport and head-scanning mechanism, which was then the industry-preferred setup.

In the early 1990's, a new HDTV recorder using Type D-1 tape cassettes was developed through the joint activities of Toshiba and Broadcast Television Systems (the successor to Bosch–Fernseh and now a Philips Company). The developers stated that this equipment was the first recorder designed not just as a video recorder but as a data stream recorder capable of accepting many different types of input data, including HDTV signals conforming to U.S., European, and Japanese standards. It is also intended for the recording of high-resolution motion picture programs, imagery, and instrumentation data.

By 1995 the Toshiba/BTS format had become the SMPTE D-6 standard.

Compressed Data Recording of an HDTV Signal

The desire to use compression for HDTV recording comes not only from the obvious incentive of media usage savings, but also from the advantages of being able to design the recorder to be small and lightweight, thus increasing its portability. As an extension of its 0.5-inch D-3 recorder design and manufacturing efforts, Matsushita in 1995 had developed a double data rate version D-3, later designated as the SMPTE Type D-5. The D-5 recorder, at the data rate of 300 Mb/s, is a useful platform for several applications. The D-5 recorder records an HDTV signal through a data compressor with a compression ratio of 4.

In late 1996 Sony Corporation completed the development of an extended capability Digital Betacam that records an HDTV signal using a 6:1 compression ratio. NHK has adapted both Matsushita and Sony recorders for its expanding Hi-Vision program production operations.

DV, DIGITAL VIDEO

DV, Universal Digital Recording Format for Consumers and Professionals

It was always assumed that the day of digital video recording would eventually come to the consumer market, replacing analog VHS for movie playback and Hi-8 and VHS-C for amateur filmmaking.

In late 1993, through an organization called the HD–Digital VCR Conference, video recorder manufactures from Japan, the United States, and Europe established an engineering project to define the technical parameters for the next generation of consumer video recorders. The parameters were to be defined for standard definition NTSC and PAL television standards, as well as for the forthcoming high-definition television standards for the three participating regions. By the end of 1994, the basic technical specifications were finalized. Participants submitted prototypes to the coordinating body for specification compliance and performance verification. The outcome of the activity is the recording format called DV. The format specifications have been submitted to the International Electrotechnical Commission and have become the IEC-1834 standard.

DV, which uses 6.35 mm metal evaporated tape in two cassette sizes, operates at 25 Mb/s for standard definition TV and 50 Mb/s for HDTV. Camcorders based on the DV format entered the market early in 1996 in Japan, and later in the year in the United States and Europe. Despite its high cost—approximately three times the cost of a high-end analog counterpart—the DV camcorder was an instant market success. Its extremely small physical size and the outstanding image quality were the two features attracting consumers.

When the quality of a consumer product reaches a certain high level, it begins to enter the professional market. When the price of a professional product becomes low enough, it becomes a consumer product. An outstanding example is the 35 mm film camera, which is the universal format now among professionals as well as amateurs.

The DV format is fast reaching that enviable position. Features attractive to amateurs are equally useful to a segment of the broadcasting industry, such as electronic news gathering. Matsushita and Sony are producing professional versions of the DV camcorder and video editing recorder under the respective trade names of DVCPRO and DVCAM. DVCPRO, which became SMPTE format D-7 in 1997, uses a high-performance, dual-coat metal particle tape as its standard recording medium.

In 1996 a data streamer based on the DV data format and transport mechanism was jointly proposed by Exabyte Corporation of the United States and Toshiba Corporation of Japan.

Broadcast-Quality VHS Derivative with DV-Based Data Compression

The DV format, being intended for consumer applications, uses a subset, 4:1:1 configuration of the original 4:2:2 component video format, as its recorder input signal, accepting a reduction in the chrominance bandwidth. Capitalizing on the availability of the double-data-rate DV format intended for HDTV application, a true professional-quality recording data format using the full-bandwidth 4:2:2 configuration was proposed in 1996.

One of the proponents of the 4:2:2 DV recording format, Victor Company of Japan (JVC), had designed its recording system based on the VHS, 0.5-inch tape transport and head-scanning configuration. The full-bandwidth digital component recording format is currently undergoing format standardization activity in SMPTE. The format will be known as Type D-9.

The DV-based video data compression algorithm, which compresses 4:1:1 data input to a 25 Mb/s data stream, and 4:2:2 input to a 50 Mb/s data stream, is now being considered by a joined EBU/SMPTE task force as a standard digital video data interconnection format to be used by the broadcast industry. This is a remarkable growth from a consumer product.

D-VHS, THE LATEST MEMBER OF THE VHS FAMILY

The VHS system has been the mainstay of consumer recorders for the past 20 years. Its family of products includes VHS, VHS-HF, VHS-HQ, and S-VHS. The professional digital component video recorder with 4:2:2 sampling configuration, Digital-S, is also a derivative of the basic system.

To meet a specific market requirement for recording directly from satellite signals, JVC introduced a new VHS format in 1995. Direct broadcasting from satellites is carried in the form of an MPEG-encoded compressed digital bit stream. A component video format is used; therefore if the signal was decoded after it had been encoded into a composite form, there is a significant degree of quality degradation on playback.

The latest VHS format, Digital VHS, or D-VHS, is a bit stream recorder operating at 19.1 Mb/s for standard definition television, or at 38.2 Mb/s for HDTV. It records the satellite bit stream as it is received from the DTH (direct to home) receiver, with the addition of an error correction overhead. To facilitate interconnections with other consumer electronics products, it is provided with an IEEE-1394 interface port.

With the total number of DTH satellite broadcast subscribers reaching the 5 million mark in the United States alone (as of April 1997), and DTH services expected to grow around the world, the D-VHS should find a sizable market in years to come.

DISK RECORDERS FOR DIGITAL VIDEO APPLICATIONS

The most labor-intensive, expensive aspect of television program production is the editing operation. The transport mechanism of a video tape recorder, with its tape reel ballistics, makes the precise synchronization of several tape recorders during editing extremely time-consuming. A genuine need for a fast-access video data source has existed since day one of tape editing operations.

Starting in the early 1980s, teleproduction operators began experimenting with computer magnetic disk memory as the video source because of its fast access time. However, at that time, the high acquisition cost, coupled with the relatively small data holding capacity, restricted the use of computer hard disks to a small segment of the industry.

In recent years, magnetic disk memories have dramatically decreased in price and increased in capacity, and several video equipment manufacturers have developed digital video disk recorders for high-end editing operators. Intended for online editing, a typical disk recorder stores a high quality 4:2:2, or 4:4:4 digital component signal for 30 to 60 seconds. The recorder is a combination of a standard computer disk memory equipped with a specialized video input/output port and an editing controller interface.

Another important application of disk memory is for the television news operations. Timely on-air presentation of news material is one of the most important parts of television broadcasting operations. Gathering the raw news material, and editing the contents to fit within an allotted time space, must be done in the shortest possible time. The raw news footage, stored in ENG recorder tape cassettes, is generally transferred onto optical or magnetic disk memory as soon as it is received by the station, or moved into the on-location vehicle that houses the editing suite. The disk players are the material source for the editing operation. The editing operation, done in a nonsequential manner to save time, is often referred to as nonlinear editing. Low-cost, high-capacity personal computer hard disk drives have become the ideal fast-access source not only for news operations but for all nonlinear program production.

Elimination of the time required to transfer news footage from the ENG camcorder tape cassette to disk memory is possible if the ENG recorder uses a disk memory instead of a tape cassette. The solution, proposed jointly by Avid Technology of the United States and Ikegami Tsushinki of Japan, combines a television camera with an IBM 2.5-inch hard disk drive. The drive is removable after recording and can be inserted into the editing system. The product entered the market in 1996–1997. A long-term solution, however, may be to use a removable optical or magnetic disk memory instead of a relatively bulky and expensive hard disk package, provided the technology advances to the level at which the disk can offer the needed capacity and transfer rate.

The third application area for the video disk memory is as a video storage system for various server operations. The concept of a separate and nonattached video storage system, controlled by a service controller/computer to meet several simultaneous but independent requests for the video material, is a highly attractive solution for many applications, including television station operation, program productions, video-on-demand services, and medical or remote sensor imagery dissemination services.

The architecture of the video or television material storage system for these services is similar to that of a large hierarchical computer data storage system. In the top layer of the storage system, where a fast access time is essential, the material is stored in the disk memory, or in the solid state memory when and if this option becomes affordable. For middle-layer applications, either a magnetic disk drive or a high-capacity optical/magneto-optical disk memory is used. Magnetic tape remains the means for mass data storage, forming the bottom layer of the hierarchical storage system.

LOOKING INTO THE FUTURE

Since the invention of the quadruplex recording in 1956, the technology of magnetic tape video recording has made astounding progress. The original quad recorder, with a monochrome video playback signal of 3 MHz bandwidth, required approximately 1200 μm^2 of tape area for the equivalent of 1 bit of data. The latest digital video tape recorder, the DVC in the long recording mode, needs only 1.6 μm^2 for 1 bit. The areal recording density improvement is nearly 750:1. The progress in volumetric recording density, which is the real measure of recording system efficiency, is approximately 3500:1. The longitudinal recording wavelength improvement may be nearing its practical limit at about 0.4 μm. The DVC operates at a wavelength of 0.488 μm. Work on the perpendicular magnetic recording (PMR), which promises to shorten the wavelength by a factor of several to one, is still going on. We may see a high-density tape recorder utilizing PMR within a few years.

The limitation on the track pitch using a flexible medium such as a magnetic tape is also a real technical hurdle for the recorder development engineers. The obstacle is not just the need to be able to place the head on the recorded track. It is the elastic, and temperature/humidity/tension-sensitive nature of the tape base film, on which the head must travel for an extended length. Engineers believe that another 2:1 reduction in the track pitch from the level of narrow track DVC to 5–3 μm is possible. It is unlikely, however, that the track pitch of a tape system will reach the level of the hard disk system in the same time period.

Analog video recording had always been the forerunner of high-density magnetic recording. The fundamental technology of high-density video recording, such as the ferrite head material, the laminated metal head construction technique, the guard band free azimuthal track concept, and metal particle and metal evaporated tape, have made significant contributions to the advancement of the overall data recording technology.

Digital video recording is a logical and natural extension of high-density analog recording. The technology will furnish cost-effective solutions to all television, imagery, and multimedia recording requirements in the decades to come.

REFERENCES

Digital Technology Development

Black, H. S., et al., "Pulse Code Modulation," *Transactions of the American Institute of Electrical Engineers,* 895–899 (1947).

Goodall, W. M., "Television by Pulse Code Modulation," *Bell Systems Technical Journal, 30,* 33–49 (1951).

Kahn, David, *The Code Breakers*, Weidenfeld and Nicolson, London, 1967.

Reeves, A. H., et al., "The 25th Anniversary of Pulse Code Modulation," *IEEE Spectrum*, 56–63, May 1965.

Digital Video Recorder Development

Baldwin, J. L. E., "Digital Television Recording," presented at the *Video and Data Recording Conference*: *IERE Conference Proceedings*, No. 26, 67–70 (1973).

Bellis, F. A., "An Experimental Digital Television Recorder," *BBC Research Department Report,* RD 1976/77, February 1977.

Ginsburg, C. P., "Report of the SMPTE Digital Television Study Group," *J. SMPTE, 85,*141–145 (1976).

Lemoine, M., et al., "Digital Video Recording—A Progress Report," *Digital Video, 2,* 139–145 (1979).

Yokoyama, K., et al., "PCM Video Recording Using a Rotating Magnetic Sheet," *NHK Laboratories Note*, Serial No. 221, December 1977.

Digital Video Recorders: Products

Engberg, E., et al., "The Composite Digital Format and Its Applications," *J. SMPTE, 96,* 934 (1987).

Oba, Y., et al., "New 8-14 Coding for High Density Recording," *J. SMPTE, 102,* 114–118 (1993).

Remley, F. M., "Introduction—The 4:2:2 DVTR Television Recording Format: Organizing the Work of Standardization," introduction to *The 4:2:2 Digital Video Recorder,* Stephen Gregory, Pentech Press, London, 1988.

Schiffler, W., et al., "A Digital 1.2 Gbit/sec VCR for a Universal Recording Format of HD Image Data," *J. SMPTE, 103,* 439–443 (1994).

Perpendicular Magnetic Recording

Taguchi, R., E. Miyashita, K. Kamijo, Y. Yoneda, J. Numazawa, "The Read/Write Characteristics of Co-Cr-Ta/Ni-Fe Double Layer Tape for Digital Video Recording," *Journal of the Magnetics Society of Japan,* **21**, (1997)

Standardization and General Reading

Remley, F. M., "International Recording Standards, A Brief History of IEC TC 60," *NML BITS,* **6,** 7–9 (1996).

Sadashige, K., "Transition to Digital Recording: An Emerging Trend Influencing All Signal Recording Applications," *J. SMPTE,* **96**, 1073 (1987).

Sadashige, K., "SMPTE's Contribution to Video, Audio, and Data Recording Format Standardization," *NML BITS*, **6**, 3–7 (1996).

Weiss, M., et al., "Putting Together the SMPTE Demonstrations of Component-Coded Digital Video, San Francisco, 1981," *J. SMPTE,* **90,** 926–939 (1981).

15 Capturing Data Magnetically

James E. Monson

STORING DATA

Given the success of early magnetic recorders in storing voices and pictures, it was natural for the technology to be applied to capturing other forms of information—broadly defined as data. The applications were to take many forms. Continuous-loop recorders in remote locations waiting for seismic or atmospheric disturbances to capture, monitoring of power distribution lines to record details of blackouts, and the "black boxes" used in airplanes are just a few examples. The development of the digital computer generated new requirements for data storage, leading to an impressive array of technologies based on storing information magnetically. This chapter supplies the context for the development of the technologies presented in the succeeding chapters.

Computer Applications

Early computing machines could perform complex tasks dictated by the "wiring" or hardware designed into them to enable particular operations. If a new application was desired, rewiring of the computer or often a complete redesign was necessary. The widespread adoption of the John von Neumann, or stored program digital computer in the 1950s changed the basic computer configuration, making it much more flexible and generating an ever increasing demand for memory or storage. Figure 15–1 shows a block diagram of a stored program computer. The processing unit is designed to be very simple, yet powerful. It executes a series of instructions that are fetched from the stored program part of the

Memory

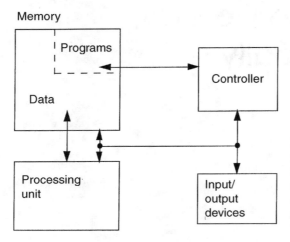

Figure 15–1   Block diagram of a stored program computer. The controller fetches instructions from the program part of the memory. These are executed by the processing unit. When data are required to execute an instruction, they are fetched from the data part of memory.

memory, under the guidance of the controller. The instructions use data stored in the data part of the memory. For example, a stored banking program might run at the end of each day, updating accounts with all transactions entered that day. The same computer could next run a different stored program on the same data to generate bank statements for a set of customers. This versatility of the stored program computer has led to its extraordinary development, and to much of the rise in magnetic recording digital data storage.

Other Applications

Apart from computer applications, magnetic recording is used to acquire and store data in many other situations. The development of large test ranges for aircraft and missile flight evaluations led to the requirement for recorders capable of storing telemetered data that arrived at very high rates. This technology became known as instrumentation recording, a term used broadly to include all forms of recording signals that represent physical phenomena varying in time. Examples are the engine temperature, oil pressure, electrical system voltage, current, and frequency characterizing aircraft instrument behavior during a flight. Early instrumentation recorders used the analog, linear, biased mode of recording of audio recorders. The time-varying signal was faithfully replicated for storage, much as the fluctuation of the needle on an oil pressure gauge was observed by a human pilot. Later, as the requirements for higher data rates developed, the video recorder technique of frequency modulation was used to increase bandwidth. Often, several measurement channels were stacked or multiplexed together to increase the speed at which information could be handled. As with audio and now, video recording, instrumentation recording uses digital recording techniques to realize extremely high performance data recording systems.

At what might be regarded the low end of the technology spectrum, credit cards and subway tickets are important applications of digital recording. Stripes of magnetic recording media contain digital data representations of the card holder's name, account number, and expiration date, or in the case of tickets, fare segments remaining. The magnetic stripes are similar to audio tape, but the digital storage format is designed for low recording density, to be very robust under adverse conditions for reading the stored information.

DATA STORAGE REQUIREMENTS

The information explosion has led to many diverse applications for data storage or memory. Each of these has its own particular requirements, which often conflict with one another. Outlined below are some of the requirements that have impacted the development of data storage and memory technologies.

Short-Term Versus Long-Term Memory

Just as in our daily lives, we distinguish between our short-term and long-term memory processes, so it is with data storage applications. For example, intermediate results in the pocket calculator addition and division to allocate shares of a luncheon bill must be stored until everyone agrees on the outcome, but after the bill is paid, few are interested. In most calculators, the data are lost when the calculator is switched off. Such a memory is said to be volatile. On the other hand, a memory using magnetic recording is inherently nonvolatile. It continues to store the information after power is removed from the recorder. The requirement for long-term, nonvolatile memory is important in many applications such as billing information in the automatic billing system of a telephone network. Magnetic storage retains the information in case of an unwanted power outage, while a volatile memory must be protected with a backup power system to prevent loss of data.

Transfer Rate, Speed, and Access Time

In characterizing memory systems, a distinction is made between the concepts of data transfer rate (bits per second), and access time. Transfer rate is limited by the time taken to write or read data in a memory. The inherent switching times between one memory state and another limit the speed of the write and read processes. The fastest switching in computer memories occurs when a semiconductor device in a random access memory is switched from off to on. There is always a trade-off between speed and power. Applying more power to a device can cause it to change states faster. With proper design, magnetic memories can change states extremely rapidly, but still more slowly than semiconductor-based memories.

Access time is the time necessary to reach the location in memory that stores the desired data. Semiconductor memories and magnetic core memories are laid out on grids, with each data location lying at the intersection of two grid lines. Given the address of the data location, the computer can access it for reading and writing by selecting the two lines forming the intersection. This is a very fast electronic process, lasting only a few cycles of the computer's clock. On the other hand, the most commonly used magnetic storage devices (drums, disks, and tapes) all use mechanical positioning to access the stored data. Consequently, their access times are much slower than those of semiconductor memories. Among magnetic devices, there is a further distinction between serial access and direct access storage. In serial access, such as in a tape drive, data are stored sequentially along the tape, which is rolled up on a reel. Consequently, if the desired data are located near the hub, or end, of the reel, a large amount of the tape must be unwound before the data can be accessed. Disks and drums have a two-dimensional layout of the data, usually in the form of rotating circular tracks. After a movable arm has accessed the desired track, the data of interest will be at most one revolution of the track away.

Storage Density

In all data storage technologies, the storage density, defined as the information stored per unit area or volume, has increased dramatically over the last 40 years. In rigid disk technology, the increase in density per unit area of disk, or areal density, has been roughly a million times. Almost all this progress has resulted from scaling down all the dimensions—head, magnetic medium, and head-to-medium spacing. Many uses require high density, including laptop computers with significant capability, large databases in mainframe computers, and aerospace operations, where size and weight are at a premium. On magnetic recording surfaces, areal density is the product of linear bit density and track density. The amount of digital information, or bits, per unit length, is the linear density. Over the years, gains in both linear and track densities have contributed to the increases in areal density. The region in which a bit is stored has remained a skinny rectangle, between 10 and 20 times wider across the track than along the track.

A high volumetric storage density is desirable when large amounts of data must be stored. Tape systems are the leaders in data stored per unit volume. The tape's thin substrate means that a significant fraction of the volume of a reel of tape contains stored data. In a disk system, density along the third dimension is related to the spacing of the stacks of disks. This density is less than 20 disk surfaces per inch.

Reliability

An ideal data storage system would always function as designed and never generate an error in the data, whether reading or writing. In practice, the deviations from this ideal are expressed in terms of the reliability of a system. Reliability has two aspects: the proper operation of the system and the generation of errors by a properly operating system. A system may fail to operate because of a failure of one of its key components. A capstan motor may fail, making a tape recorder inoperable. A head flying over a disk may "crash," coming into hard contact with the magnetic surface to destroy stored data and wear out the head–disk interface. Reliability specialists integrate the failure rates of individual components to obtain measures of the overall system reliability. Depending on the reliability requirements for a particular application, redundant systems or subsystems can be designed to create reliable systems from somewhat unreliable parts. Space vehicles, communications networks, and online reservation systems typically require very high reliability of operation and use redundant components and backup resources.

Requirements for error rate control also depend on the use of a particular data storage system. In general, error rates must be much lower than those encountered in a typical audio or video recorder. Advances in integrated circuit complexity have allowed redundancy techniques such as error-correcting codes to be employed to reduce errors from many sources. The improved performance is well worth the additional cost and complexity of adding redundancy.

Operating Environments

Often a product under development that performs as designed under laboratory conditions will fail in the field because it cannot meet specifications in certain environments. A magnetic medium may have very desirable properties, but if these are too sensitive to temperature or humidity, some environments are ruled out for that medium. As computers moved out of very carefully controlled surroundings in large computing centers to homes, offices, and shops all over the world, design efforts to meet environmental requirements intensified. Temperature, humidity, salt water, trace amounts of corrosive gases, shock, and vibration must all be accounted for.

Cost

Cost is a major consideration in the adoption of any technology. In data storage, the measure is cost per unit of data. Cost per bit of magnetic storage has decreased dramatically as storage density has increased. Semiconductor memory technology has experienced a similar trend, so that the ratio of semiconductor memory cost to magnetic disk cost has remained roughly 100:1. This large cost differential has led to a hierarchical organization of computer memory that integrates the various technologies to obtain an optimum cost–performance outcome. Disk and tape memory cost is now so low that the maintenance cost of memory in terms of backing up information, cataloguing, and servicing memory systems is far higher than the installed cost of the storage systems themselves. The steadily decreasing cost of magnetic disks and tapes has been an important factor in shutting out competing technologies that have challenged for the storage market over the years.

Removability and Portability

Many storage applications require the ability to remove a storage module, such as a disk or tape, and to replace it with a blank, or with another module containing data. Removable modules may be used on the same machine or have the capability to be used on other equipment. The latter property, portability, is illustrated by the story of a financial institution with two large data processing centers—one in northern California and one in southern California. At the end of each business day, a set of reels containing taped backups of the day's transactions was loaded into an armored truck at each site. The trucks then headed toward each other, meeting at a destination in the middle of California. After an exchange of tapes, each truck returned to its starting point. Portability of the tapes was essential, and the system achieved a geographical redundancy of data as well. A less exotic example of portability is the transfer of data or programs between personal computers using floppy disks. In spite of industry standards to ensure portability, data are sometimes not transferable to a new computer.

If removability is not required, the box containing the recording medium, heads, and mechanical positioner can be sealed to create a controlled "clean room" environment. This is the approach that was used in the Winchester disk technology, and it led to higher performance than could have been achieved if the disk had been removable.

Archivability

Archivability is the ability to store data for very long periods of time, in effect, making a permanent record. Magnetic records of data are relatively stable, but they do decay over time, both magnetically and physically, in accordance with theory and experimental evidence. Other practical details of housing the records are just as important. Most archival records are on tape, and these must be maintained in accordance with the properties of the magnetic media and substrate films involved. Another consideration is the maintenance of the equipment used for recording and playback. These systems have undergone constant improvement and modification over the years, so that today's equipment may not be capable of reading an archival record recorded years earlier. In fact, this incompatibility is one of the most common factors limiting the usable lifetime of archival data.

EARLY COMPUTER STORAGE SYSTEMS

Before magnetic recording products became the primary choice for data storage in computer systems, developers tried a number of technologies over the years to satisfy some combination of the requirements of fast access time, high reliability, low cost, and nonvolatility.

In the early 1940s, data were stored on punched paper tape. This medium had the advantages of low cost, nonvolatility, and relatively good reliability. On the other hand, access time to the data was very long because of the slow serial access required to get to a specific data record.

Faster storage devices developed during the 1940s used a variety of new technologies. Acoustical delay lines made from solid state materials like quartz or glass could store for about a millisecond a stream of bits of reasonable length. The devices had a long path over which the bits propagated, reflecting off the facets of the quartz crystal. These memories were volatile, but because of the low acoustic attenuation in the quartz, the memory retention was long enough to make them usable. An alternative approach was the Williams cathode ray tube, storing data in an array of several thousand bits on the face of the tube. The data could be accessed 10 times faster than was possible in the acoustic delay line. The Williams tubes were also volatile, but the refresh rate was significantly better than that of the delay lines. Reliability was not high, but no worse than for the vacuum tubes used in other parts of the computer.

The late 1940s saw the introduction of the first magnetic memory device. It consisted of an array of magnetic toroids or cores, each core storing one bit of data. The cores were threaded with wires in a way that allowed random access to each core for writing and reading. The magnetic storage was nonvolatile, but the information was erased in the read process, necessitating the immediate rewriting of the read bits. This core memory technology was reliable, fast, and relatively inexpensive. It was successful in displacing the earlier technologies, and its performance improved significantly with the introduction of new magnetic materials and manufacturing techniques. Magnetic ferrite materials were developed for use as cores, and by the early 1950s, three-dimensional core arrays were being built with capacities of thousands of bits. These could be accessed reliably in microseconds.

Although ferrite cores were very successful, computer applications expanded rapidly, and many uses required memory capacities far in excess of what core memories could meet at a reasonable price. Computer system designers had to consider cheaper storage technologies, even if they were significantly slower than cores. Magnetic drums offered just this combination of price and speed. Designers combined drums and cores in a system that automatically loaded data from the drums into the faster core memory when requested by the CPU. During this fetch of data, the computer stopped its normal operations. Processing time was slowed, but many more tasks were possible because of the greatly expanded memory capacity, referred to as "virtual" memory because it did not have the same short access time as the main core memory. Multiprogramming techniques allowed the CPU to execute parts of programs while waiting for transfer of requested data from the slower storage device into the fast main memory. The components of a storage system became organized into a hierarchy linking expensive, fast, small-capacity memory to inexpensive, slower, large-capacity data storage devices.

Initially, magnetic drums were the preferred large-capacity storage device, but by the 1960s, magnetic disks had become dominant. Although slower than drums, they were much cheaper and had potential for greater capacity. At the same time, semiconductor memories were replacing cores as main memory devices, and the stage was set for today's memory hierarchy organization. This configuration has persisted for so long because the costs of semiconductor storage and magnetic disk storage have decreased dramatically at roughly the same rate. Hence, both technologies have survived the challenges from other storage technology developments over the years, although at the time they promised the winning combination of very low cost and very fast access.

THE MEMORY HIERARCHY

The memory hierarchy discussed above exploits the advantages of the various memory technologies, while mitigating the less desirable features. Figure 15–2 illustrates the concept in the form of a triangle. As one moves up the triangle, access rate increases. Access rate is usually proportional to cost per unit of memory. The width of the triangle at any section corresponds to memory capacity. Programs and data are freely transferred from one level of memory to another, exploiting the idea of the stored program computer.

The main memory, usually called random access memory, or RAM, contains the programs and data the computer processor needs to execute its tasks. Because the computer is constantly accessing this memory at its very fast operating speeds, the main memory must have high speed and very short access time. These requirements today dictate the use of semiconductor technology for the main

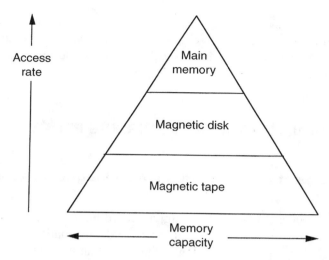

Figure 15–2 The memory hierarchy in terms of access time against memory capacity. Access time includes addressing the data location, writing, and reading.

memory, with access times of 50 to 100 ns. Because of the relatively high cost of semiconductor memory, the capacity of the main memory is designed to be as small as possible without compromising overall computer performance. The activities or transactions taking place in the main memory are often stored for only a short time. Consequently, the nonvolatility of semiconductor storage is not a problem. Data that must be stored for long periods are readily transferred down the hierarchy to magnetic disk or tape, but this must be done at rates slower than the basic computer operating speed.

At the second level down are disk memories. Access times are about 10 ms, much longer than for main memory, but the ability to access data directly places disks in the hierarchy above sequentially accessed magnetic tape, which takes minutes to traverse the length of the tape. Both disk and tape are nonvolatile and can back up critical data generated in the main memory. Low cost and high storage densities make practical the large capacities of the disk and tape portions of memory. Disk and tape levels can each have sections with removable and portable memory modules—floppy disks, reels, or cassettes, for example. Archivability, if required, is usually accomplished with magnetic tape.

There has always been a significant difference in access time between semiconductor main memory and disk. This "access gap" drove the development of many techniques, and in some cases, technologies to solve this problem by "filling the gap." The most successful approaches have entailed the use of layers of temporary buffer memories, caches, to store data from disk that will be accessed often in the execution of the computer program at hand. These buffer RAM chips are employed both by the computer's central processing unit and controllers on the disk drives themselves. The staging of data among main memory, cache, and disk is managed by software control that can be quite sophisticated. At the moment, competing technologies to fill the "access gap," such as magnetic bubble memories, have been squeezed out by the vigorously growing semiconductor and magnetic disk technologies.

THE DIGITAL MAGNETIC RECORDING PROCESS

Storage of Bits of Information on Magnetic Media

As in analog recording, digital recording uses the hysteretic properties of so-called hard magnetic materials to record information. In digital recording, the process is simpler than analog in the sense that there are only two orientations for the magnetic dipoles (magnetization) in the medium—either forward or reverse along the track. The track is divided along its length into data storage locations or bit cells. The conventional way to store data is to have a reversal of the direction of magnetization in a bit cell correspond to a data 1. No reversal (i.e., uniform magnetization throughout the cell) corresponds to a data 0. The bit cell boundaries are not laid out physically along the track, but are determined by timing of when the write and read heads pass over a given location on the track. For modern rigid disks, the magnetic recording medium is a continuous "featureless" cobalt alloy film deposited on a substrate, typically of aluminum or glass.

Writing

Writing is the process of imparting the correct orientation of the magnetization in the magnetic medium to store the desired data. The write head is simply an electromagnet with a small gap in its core. To write a data 1 at a desired storage location, one reverses the direction of the coil current at the time corresponding to the location of the bit cell center opposite the head gap. This current reversal writes the desired magnetization reversal (often referred to as a magnetization transition) at the correct location. Usually, the storage device is operated so that new information to be stored can be written right over the old information. The head writing field is strong enough to remove effectively all traces of previously written data.

Ideally, a written magnetization reversal or transition extends over an infinitesimally short distance along the track. In practice, the transition occupies a finite length, which places a lower limit on the size of a bit cell. The resulting finite transition length depends on the magnetic properties of the magnetic medium, its thickness, the head design, and the spacing between the head and medium.

Reading

The magnetic poles at the transition locations produce magnetic flux. Data 1s can be read back by sensing this flux. A "no flux sensed" condition corresponds to a data 0. In early storage systems, reading and writing were almost always done with the same inductive head. In the reading process, the change in flux in the head core as it passes over a transition on the medium induces, by Faraday's law, a voltage pulse across the head coil.

Many other flux sensors can be used to read stored data. An increasingly popular device is the magnetoresistive sensor. This is a thin film of magnetic material whose resistance changes as a function of the magnetic field (flux) incident on it. In operation, a current is passed through the film, and the changes in voltage across the film correspond to the pulses in magnetic flux from passing magnetization transitions. Unlike an inductive read head, the magnetoresistive sensor output is independent of the head–medium velocity. Read heads using such sensors can improve system performance significantly and have led to a dual-head configuration—a conventional head with gapped core and coil for writing and a magnetoresistance head for reading.

Head–Medium Interface

The nature of the fringing flux from a magnetization transition is such that the width of the flux pattern read by any sensor increases rapidly as the sensor moves away from the surface of the storage medium. This is a property inherent in magnetic recording, making the readback pulse width, or resolving power, critically dependent on head-to-medium spacing. The write process is also sensitive to head–medium spacing, making this parameter perhaps the most important one determining overall recording system performance. The understanding and control of the head–medium interface has been critical to the continued development and improvement of magnetic data storage. A number of technical disciplines are involved: mechanics, aerodynamics, gas lubrication theory, friction, wear, and sophisticated measurement techniques are all significant. Friction, lubrication, and wear are often grouped under the general subject of tribology.

Tape and flexible disk systems operate so that heads are nominally in contact with the medium, implying that the head–medium spacing is zero. In fact, both head and medium have a surface roughness that prevents perfect contact. In addition, there may be asperities, primarily on the medium surface. These tiny defects cause momentary separation between head and medium, with loss of data or "dropouts." In contact recording systems, the contour of the head surface and the surface smoothness of the medium are the important factors in maintaining the desired interface conditions as the medium moves by the head at high velocity.

In rigid disk systems, the head and medium are not in contact, but the head flies over the medium at a small controlled "flying height." In most designs, the head is mounted on a "slider," a small, sled-like structure with rails. Sliders are designed to form an air bearing that gives the lift force to keep the slider-mounted head flying at the desired head–medium spacing. Properly designed, the slider follows any irregularities in the disk surface and maintains spacing as it moves outward in radius to fly at a higher linear velocity. Tribological issues, as well as contaminants in the environment, are of constant concern.

Tracking and Accessing Data Mechanically

In semiconductor memories, and some other memory technologies, bit cells or other data storage regions are defined by physically depositing a pattern on the area that will store the data. For magnetic disks, drums, and tapes, the storage medium is homogeneous, and storage locations have to be identified by a track number and a coordinate along the track. The location of a bit cell along the track is obtained from the signals recorded by the head. These are coded in such a way that the information is stored and timing signals generated so that a synchronized counter can identify the particular bit cell along the track of interest at the time of readback.

The track is defined by the writing process itself, which is capable of very high resolution. The magnetic medium starts out as if it were a blank sheet of paper. Tracks on the magnetic storage medium may be defined in a number of ways, but the basic principle is to record a set of control or marking signals to identify each track. On a disk, the tracks are laid out in concentric circles, as shown in Figure 15–3. When the head is reading or writing data on a particular

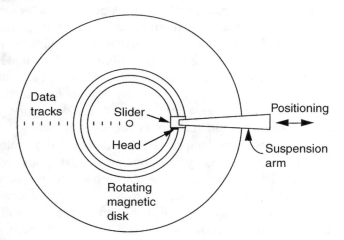

Figure 15–3   Concentric circular tracks on a disk: a positioner moves the slider containing the head to either follow a track or seek a new track.

track, it is very important that it track or align itself on the center of the track. Accurate tracking ensures that the maximum signal strength is read back from the track and minimum interference is sensed from data signals on adjacent tracks. The recorded control signals are designed to produce an error signal when the head moves off center. A control system or servomechanism uses this error signal to position the head back on track center.

When the computer needs data that are stored on a different track, the tracking control system changes mode of operation from tracking to accessing a new track, an operation referred to as seeking. The control system generates signals to control the positioning motor or actuator in a way that minimizes the access or seek time between tracks. The commands are often given in terms of an offset from the present track location (e.g.,"move in 10 tracks" or "move out 3 tracks"). As the positioner moves the slider in or out, the head reads a track control signal each time that it crosses a track. The controller counts how many tracks are left until the positioner must stop, and modifies the control signal to the actuator accordingly.

Storing and Retrieving Data from Magnetic Memory

Although not so daunting perhaps as sending and receiving data from Mars, the procedures for storing and retrieving data from magnetic memory are often highly sophisticated. The data information handling portion of a magnetic storage system is called the magnetic recording channel, analogous to a communications channel. We have already seen how data are written by reversing the sense of magnetization in a bit cell, and then read by sensing the voltage pulse produced by the magnetization transition. These basic operations are affected by several important physical parameters and processes, such as recording medium demagnetization, which spreads the width of the transition, and head–medium spacing, which broadens the readback pulse. Another physical process critical for recording channel design is the creation of various sorts of noise and interference.

Noise is created by random fluctuations in physical variables. For example, the bulk magnetization in a magnetic medium is the average of many small magnetic regions, or domains, making up the medium. These domains have a certain randomness in orientation that can impart an uncertainty to the location of a magnetic transition. When the computer is trying to sense the presence of a transition in a bit cell, to make a decision of 1 or 0, the "noisiness" or uncertainty in the transition location may cause an incorrect decision to be made, hence increasing the bit error rate. Other important sources of noise are in read heads and the electronic circuits used to process the readback voltage signals. Interference sources are not, strictly speaking, random. They consist of man-made signals other than the desired one. For example, data recorded on an adjacent track could be sensed by a read head that is somewhat off its desired track. Residual signals from data written earlier and then written over with new data can also cause interference.

Given a characterization of the writing and reading processes along with the important noise and interference sources, the figure of merit for a recording channel is expressed as the signal-to-noise ratio (SNR). In data recording, signal-to-noise ratio is an extremely sensitive indicator of bit error rate. A change in SNR of 10% can correspond to a 10:1 change in bit error rate.

Magnetic recording design proceeds outward from the basic read, write, and noise processes as shown in Figure 15–4. Generally speaking, the input data cannot be directly applied to the write process for several reasons. Recall that the bit cell address along the track is obtained from a counter that counts the number of bit cells synchronously with the data. To maintain synchronism, the counter must receive a minimum number of readback pulses periodically. If the input data were written directly, long strings of data 0s, often encountered in practice, would generate no readback pulses (no magnetization transitions) to keep the counter in synchronism. This problem was recognized very early in the development of data recording. System designers invented various methods of coding the input data so that long strings of 0s were coded into strings containing enough 1s to synchronize the address counter correctly. Such codes were designated modulation or channel codes. They were also used to treat the prob-

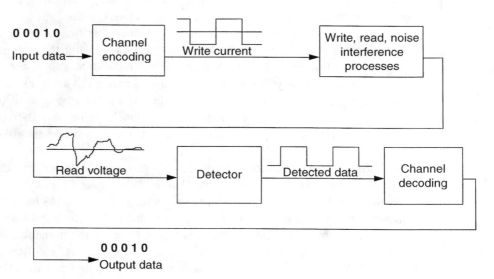

Figure 15–4   Magnetic recording channel, showing data flow through the system. Often, error-correcting coding blocks are added around this channel.

lem of interaction between two 1s spaced close together. This situation could be avoided by the proper choice of channel code.

The detector must process the readback signal along with noise and interference and make a decision whether a 1 or a 0 is stored in a particular bit cell. Early detectors simply sampled the readback signal during a small "window" in the center of the bit cell. If the voltage exceeded a certain threshold value, a 1 was declared to be received. A sample below threshold indicated a 0. This method was very amplitude-sensitive, and the peak detection technique was developed to overcome amplitude variability caused by head–medium spacing fluctuations, medium defects, and so on. If a pulse peak occurs in the detection window, there should be a zero crossing of the time derivative of the readback waveform. The derivative waveform crosses through zero for both positive and negative pulses, indicating the presence of a 1. More recently, detectors have incorporated signal shaping and decision strategies used in advanced space communications. The success of these approaches has led to the possibility of dealing with second-order noise and interference effects that were formerly neglected.

After detection, the detected signal is decoded to produce the original data from its channel-encoded version. Often a second coding and decoding process is added around the channel shown in Figure 15–4. This accomplishes error correction. Redundant bits are added to the data so that only certain patterns, out of the many possible, represent valid data. When an invalid pattern is discerned after detection and channel decoding, it is clear that an error has occurred. Within constraints, the nature of the error pattern can be used to infer the original, correct data. The bit error rate can be improved by several orders of magnitude at a reasonable cost in redundancy.

Out of the million times improvement in storage density over the last 40 years, improvements in channel and error correction coding have accounted for a factor of 10. Although small compared with the remaining improvements due to scaling of the recording process, channel enhancements have been and continue to be important.

CONCLUSION

This chapter has presented background for the description of the development of magnetic data storage products given in the chapters that follow. Starting from a brief survey of applications, we have seen how requirements developed over the years to meet the needs of various storage users. The nature of the applications led naturally to a technology framework, or storage hierarchy, where the particular properties of each magnetic technology (disk, drum, tape, etc.) can be utilized advantageously to optimize performance to meet requirements.

REFERENCES

Johnson, L. R., J. H. Palmer, and E. W. Pugh, "Strength in Storage Products," in *IBM's 360 and Early 370 Systems,* MIT Press, Cambridge, MA, 1991, pp. 220–240.

Mee, C. D., and E. D. Daniel, Eds., *Magnetic Storage Handbook*, 2nd ed. McGraw-Hill, New York, and IEEE Press, Piscataway, NJ, 1996.

Time-Life Books, Eds., *Computer Basics*, Time-Life Books, Alexandria, VA, 1989.

Time-Life Books, Eds., *Memory and Storage*, Time-Life Books, Alexandria, VA., 1990.

White, R. M., "Disk Storage Technology," *Scientific American,* August 1980, pp. 138–144.

Wood, R. W., "Magnetic Megabytes," *IEEE Spectrum,* 32, May 1990.

16 Data Storage on Drums

Sidney M. Rubens

THE DEVELOPMENT OF MAGNETIC DRUM TECHNOLOGY

During the second world war, the U.S. Navy had a classified unit called Communications Supplementary Activity—Washington, known by the memorable acronym CSAW. Based in what had been a private girls' school, the Mount Vernon Seminary, CSAW was staffed by a collection of cryptologists, mathematicians, engineers, and physicists whose job was to break Axis codes and pinpoint the movements of enemy ships by intercepting and decoding enemy radio transmissions.

The work of the group was coordinated by several supervisors, among them Commander Howard C. Engstrom and Lt. Commander William C. Norris. In civilian life, Howard Engstrom had been a professor of mathematics at Yale. Bill Norris had been a sales manager for Westinghouse, and later an electrical engineer at the U.S. Navy Bureau of Ordnance.

The group's efforts and needs for special-purpose devices were supported by the U.S. Naval Computing Machine Laboratory (NCML). This laboratory was located at the National Cash Register Company in Dayton, Ohio, and had been headed by Captain Ralph I. Meader. Through Meader, Engstrom and Norris met John E. Parker, an investment banker and a graduate of the U.S. Naval Academy, who was then president of Northwest Aeronautical Company, a manufacturer of gliders for the Army Air Force during the war. After the war, Parker was looking for a new business venture.

In the 1946–1947 period a new company, Engineering Research Associates (ERA), was formed to do contract research and develop the same sort of equipment that had been supplied to CSAW by the Computing Machine Lab. ERA initially had offices in Washington, D.C., and later in Arlington, Virginia. Its research, development, and manufacturing operations were located on the premises of Northwest Aeronautical in St. Paul, Minnesota. Parker became the president of ERA, and Engstrom, Meader, and Norris were made vice presidents.

Two CSAW veterans, John Howard and Charles Tompkins, were named directors of development and research, respectively. The company hired several more of the CSAW staff, as well as several veterans of the Computing Machine Laboratory, the Naval Ordnance Laboratory, and the Office of Naval Research. Because ERA was not incorporated in Minnesota until January 1947, the U.S. Navy Bureau of Ships contracted with Northwest Aeronautical to start, in 1946, the work that was to be performed by ERA after it was incorporated. At the same time, the Computing Machine Lab, its staff of officers and enlisted personnel, as well as its store of laboratory equipment and materials, were moved to Northwest Aeronautical. The NCML staff monitored the work to be performed by ERA.

Among the equipment brought to St. Paul was a Magnetophon tape recorder and several reels of Magnetophon Type L magnetic recording tape. The recorder was in poor shape and was never used for recording in St. Paul. Fortunately, however, the recording and reading heads were in good condition, and so were the reels of tape.

The author of this chapter had worked in the Research Division at the Naval Ordnance Laboratory on the protection of ships against magnetic mines and on other magnetic ordnance problems. His supervisor there was William Fuller Brown, who passed on a great deal of useful information about ferromagnetic measurements. The author was hired by ERA along with Howard Daniels and Robert Gutterman, who also worked on magnetic devices at the Naval Ordnance Lab.

In 1946 the director of research was C. B. Tompkins, who explained to his new staff that there was urgent need for a large-capacity, rapid-access store of digital information. The idea was to augment, and possibly replace, the punched tape or photographic film that had been used for storing alphabetical and numerical data. Tompkins suggested that we investigate the use of some form of magnetic recording on a revolving surface such as a disk, a drum, or an endless belt. Since rapid access and high-speed operation were desired, it seemed that a drum, with parallel recording and reading of digital words, would be the quickest and easiest approach to accomplish the desired goal.

Inasmuch as the Magnetophon tapes and heads were immediately available, it was decided to investigate the use of a small drum with strips of magnetic recording tape cemented to its surface. The heads would be used in a fixed, non-contact mode: that is, sufficiently displaced from the tape's surface to prevent wear of the tape or head. For this purpose, the experimental drum shown in Figure 16–1 was used. The drum, 5 inches in diameter and 1 inch long, was machined from aluminum.

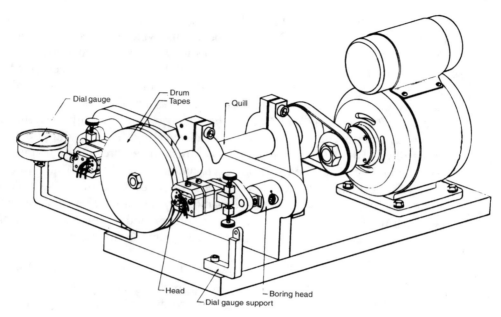

Figure 16–1   Schematic view of a variable speed magnetic tape spinner: two
tapes are glued to the surface of the 5-inch drum. [*Courtesy of
The Babbage Institute and Lockheed-Martin Corporation.*]

As shown in Figure 16–1, the drum was mounted on the end of a grinding-quill shaft . The drum's cylindrical surface was then cut to its final finish so that it did not vary from its nominal radius by more than 0.1 mil. A 1/8 hp motor, with a rated speed of 1728 rpm, was used to drive the quill shaft by means of a flat belt and a pair of interchangeable pulleys of different diameters. With this arrangement, drum surface speeds ranging from 125 to 1570 in./s were attained.

RECORDING MEDIA

As already mentioned, reels of Type L Magnetophon tape were available. Also, visits were made to Semi Begun and Otto Kornei at the Brush Development Company to obtain Brush recording/reproducing heads and also samples of the company's magnetic recording tape. The Brush 0.25-inch wide plastic tape was coated with a dispersion of acicular γ-Fe_2O_3 particles to a thickness of 0.8 mil. In addition, W. W. Wetzel, a former professor of geophysics who worked for the Navy Department, and Robert Herr, who had worked on magnetic mine countermeasure problems at the Naval Ordnance Laboratory during the war, had come to the Minnesota Mining and Manufacturing Company in St. Paul to develop magnetic recording tape. They first attempted to coat paper with a dispersion of synthetic magnetite (Fe_3O_4) particles, and several samples of such tape were obtained. Later, Wetzel provided other specimens, which were similar to the Brush tape in their pulse recording characteristics with the experimental drum.

Two tape strips, whose ends were cut diagonally, were cemented to the drum's surface as shown in Figure 16–1. One tape had a series of binary 1s recorded on it all around the drum. This served as a timing track. The second strip was used for recording and reading the signal return-to-zero pulse information. The read/write heads, one on each track, were modified recording/reproducing heads (Type BK-919), manufactured by Brush. The core of one of these heads consisted of two C-shaped sections each containing eight 0.014-inch thick laminations of Permalloy stacked to a thickness of about 0.125 inch, and wound with a coil. The two half-sections were separated by a small active front gap (in contact with the tape when used on a tape recorder), and a larger passive back gap. The inductance of this head at 1 kHz ranged from 200 to 400 mH, and the resistance was 1.5 ohms.

Several of these heads were modified by rewinding the core pieces and readjusting the width of the gaps to give optimum signal amplitude and pulse resolutions. The most satisfactory head for recording, reproducing, and erasing pulses at rates up to 30,000 pulses per second consisted of a Brush head in which the total number of turns in the winding had been reduced from 1800 to 800, and the front gap increased from 0.001 inch to 0.003 inch. The inductance of the 800-turn winding with 0.003-inch gaps was found to be 21 mH, giving a much lower time constant than the original head.

Return-to-zero recording was accomplished by first erasing the tape with a large constant direct current flowing through the head winding. Binary 1s were recorded by current pulses in a direction opposite to that of the erasing current. The absence current, or pulses of current in the direction of the erasing current, are used to record binary 0s.

Since the read/write heads were to be used in a noncontact mode, it was essential that their displacement from the recording surface of the drum be fixed and sufficient to prevent wear of either the head–core or the recording surface of the drum, but still as small as practicable to provide adequate signal intensity when the recorded information was being read. At various drum speeds, studies were made of the effect of increasing the head displacement from actual contact up to 4 mils.

Bertil Lindquist, an aeronautical engineer, suggested that the read/write heads be mounted on gimbals and allowed to fly on the boundary air layer that is next to the surface of the rotating drum. Inasmuch as the drum to be developed was to be rotated in steps, so that the drum would be at rest during the recording of a word of information, such a "flying head" would not have been suitable in this application. Years later, however, pivoted flying heads were used with large-magnetic drums to attain smaller spacing of the heads from the drums, and consequently achieve much higher recorded densities.

Adequate signal-to-noise ratios were found with a displacement of 0.002 inch between the head and drum at a drum surface speed of 400 in./s. Tape coated with γ-Fe_2O_3, supplied by the 3M Company, was used in this study. Pulses representing binary digits (bits) were completely resolved at a bit density of 80 pulses per inch. These experimental observations were the basis for the development and production of magnetic drums using fixed, noncontact heads for data storage. The experimental study was completed and reported to the navy on June 19, 1947.

John L. Hill and Robert I. Perkins were the engineers responsible for the development of the first full-scale magnetic drums that were delivered to the sponsor. These drums were made of cast aluminum and, like all future drums with fixed noncontact heads, they were surrounded by an aluminum shroud. There was a hole in the shroud at every position where a head was to be mounted. Each hole was drilled in a direction normal to the surface of the drum and threaded. Perkins designed a head-holding container that had a threaded cylindrical outer surface. This head holder could be screwed into one of the threaded holes in the shroud in the same manner that an automobile spark plug is screwed into its receptacle in the engine. An exploded picture of the head is shown in Figure 16–2.

To properly mount a head, the holder was screwed into the shroud until the active gap of the core contacted the drum's surface. Then, using a dial gauge to measure displacement, the holder was unscrewed until the desired displacement was attained. The first two drums constructed to satisfy the Bureau of Ships contract were 34 inches in diameter and about 12 inches long.

The recording surface consisted of strips of 3M ferric oxide coated Mylar tape cemented to the drum. An engineering model of this drum is shown in Figure 16–3. The drum contained a total of 25 tracks. One was a timing track for self-clocking, and the 24 others were for recording 24-bit words in a parallel mode. A gear with 5340 teeth was attached to the perimeter of the drum on one side. This gear was driven by a worm gear attached to a stepping motor for initially loading the drum with data, the drum being at rest each time a 24-bit word was recorded. The data were supplied from a punched tape (Teletype tape) whose motion was synchronized with that of the stepping motor. After loading, the gears were detached, and the drum was revolved at 600 rpm for reading and processing the data. Each data track had three heads, for recording, reading, and erasing information. The initial goal was to be able to read the data at 10,000 words per second. In fact, data were recorded at a density of 50 b/in. along each track so that, at a drum speed of 600 rpm (10 rps), a data reading rate of 50,000 words per second was achieved.

After 1948 significant improvements were made both to the recording medium and to the heads. Thus, 3M supplied a lacquer dispersion of the ferric oxide particles used to coat recording tape, and this was spray-painted on the drums made at ERA thereafter. When manganese ferrite became available, it was used

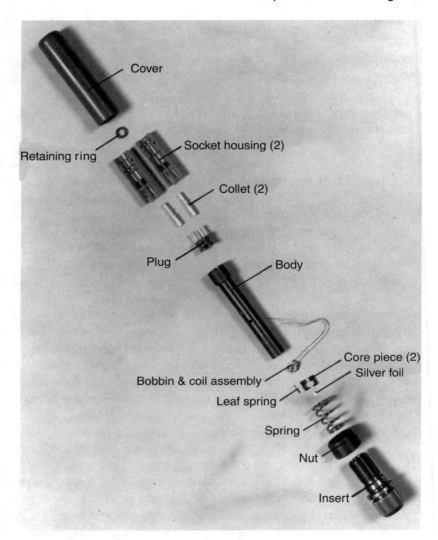

Figure 16–2 Exploded view of cylindrical head mount with differential screw arrangement. [*Courtesy of The Babbage Institute and Lockheed-Martin Corporation.*]

for fabricating head–core yokes instead of Permalloy or mu metal laminations. Because of its higher resistivity, ferrite's effective permeability at high frequencies is greater than that of metal laminations. Additionally, John Hill found that inserting a thin silver foil in the active gap of the ferrite core improved the recording efficiency of a head.

In 1948 Arnold Cohen, William Keye, and Arnold Hendrickson showed that with a suitable addressing system, a single bit of information recorded on a drum could be selectively altered without erasing or rewriting all the informa-

Figure 16–3 Two views of a 34-inch storage drum with several magnetic tapes glued to its surface. [*Courtesy of The Babbage Institute and Lockheed-Martin Corporation.*]

tion. A single binary 1 could be changed to a 0, or vice versa, without a significant increase in the background noise when the altered information was read. U.S. Patent 2,540,654 was granted to Cohen and his associates for this storage system.

The next drum development at ERA served as the internal memory for the Atlas 1 computer, which was delivered to the navy in Washington in October 1950. The drum was 8.5 inches in diameter and 14 inches long. It was sprayed with a layer of dispersed ferric oxide particles. The drum stored 16,384 24-bit words recorded at 80 b/in. along each track. The drum revolved at 3450 rpm, so that the maximum access time was 17 ms. The computer employing this drum was called the ERA 1101 (binary 13) since it was built under Task 13 of the Bureau of Ships contract.

In 1952 ERA was merged with the Remington Rand Corporation and, in 1955, Remington Rand and the Sperry Corporation merged to form Sperry Rand. Soon after, ERA, the Eckert-Mauchly division of Remington Rand, and the Norwalk, Connecticut, Laboratory of Remington Rand, were merged. The new unit was the UNIVAC division of Sperry Rand, which was managed by William Norris. From that time on, drums built in St. Paul were designated as UNIVAC 1100 series drums. They all were spray-painted with gamma ferric oxide, and used the barrel-type head mount screwed into the shroud shown in Figure 16–4.

A computer built at the ERA facility for the U.S. Air Force was known as the ERA 1102. Its internal memory was a smaller drum, 4.5 inches in diameter and 8.0 inches long. The drum stored 4096 24-bit words and rotated at 6900 rpm, so that the maximum access time was 8.7 ms. Subsequently, under the Bureau of Ships contract, a more advanced computer, designated Atlas II, was constructed using a magnetic drum to supplement an electrostatic storage tube memory. This drum, 17 inches in diameter and 10 inches long, had a sprayed ferric oxide coating. It stored 16,384 36-bit words and rotated at 1725 rpm, so that the maximum access time was 34 ms. This computer's first use by the government

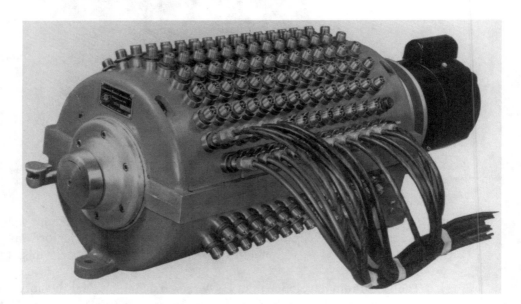

Figure 16–4   ERA Type 1100 magnetic storage drum, showing some cables attached to the head mounts. [*Courtesy of The Babbage Institute and Lockheed-Martin Corporation.*]

was in September 1953. A commercial model of this computer, using the same drum as the Atlas II, was available in the first quarter of 1954.

The ERA division sold a variety of custom-designed magnetic drums to government agencies, as well as to industrial and academic users. These drums were all of fixed, noncontact head design, but of various lengths and diameters to meet customers' needs. Most used the aluminum shroud with barrel-type head mounts screwed into holes in the shroud. Several of these are shown on Figure 16–5.

Other drums of this sort were employed to store alphabetic or numeric information expressed as binary digits in what was called the UNIVAC File computer. The drum was of the series 1100 type, but there was only one head per track. Each drum stored 180,000 characters. The average access time was 17 ms, and drums could be used in a group to store up to 1.8 million characters. The information in the file was stored as unit records, each unit record being the equivalent of a tabulator punched-card record. The first file computer system was delivered to a customer in 1956.

Figure 16–5 ERA engineers with a variety of magnetic drums. The engineers include A. Cohen, A. Hendrickson, J. Hill, F. Mullaney, and R. Perkins. [*Courtesy of The Babbage Institute and Lockheed-Martin Corporation.*]

MAGNETIC DRUMS AT HARVARD UNIVERSITY

The Harvard Computation Laboratory, directed by Howard Aiken, designed several computers. Two of these, designated as Mark III and Mark IV, contained magnetic drums. The Mark III computer, built for the U.S. Naval Proving Ground at Dahlgren, Virginia, was owned by the Navy Bureau of Ordnance. It had two aluminum magnetic drums, which were vacuum-cast by the Aluminum Corporation of America. One drum stored 200 16-decimal-digit words, plus sign, with a maximum access time of 5 ms. This computer was ready for use on January 1, 1951. The Mark IV computer was built by the Harvard Computation Laboratory and the Darmstadt Technische Hochschule in Darmstadt, Germany. It also had two drums, each 8 inches in diameter, running at 7000 rpm. One drum stored 4000 16-decimal-digital words, plus sign; the other stored 10,000 8-digit instructions. The drums revolved at 1800 rpm, so that the maximum access time was 37 ms. There were 1600 tracks, with one read/write head per track. Four tracks had two heads each. The computer was ready for use in 1952.

THE UNIVERSITY OF MANCHESTER COMPUTER

At the University of Manchester in England, magnetic drum development was begun in early 1948. A drum was needed to supplement the electrostatic storage tube (Williams tube) memory that had been under development by Professor F. C. Williams and his students for the University of Manchester computer. In February 1948 the university asked Ferranti of London to make a magnetic drum for Manchester's Department of Electrical Engineering. Previously, Thomas Kilburn, who had worked with Professor Williams at the Royal Radar Establishment, and was also one of Williams's doctoral candidates at Manchester, had used a drum that had been "machined rather roughly" and nickel-plated in a nearby shop. Thus there was definitely an effort to have a drum for the Manchester computer dating from February 1948.

The people at Manchester had been in contact with A. D. Booth at Birbeck College, University of London, about the design of reading/writing heads for magnetic storage systems. As part of his work, Booth produced the first thorough analysis of recording head fields. In February 1951, the serial mode computer data store, including the Ferranti drum, was completed, and a demonstration at the University of Manchester showed that the magnetic drum and Williams tube stores were indeed complementary.

The drum development by Ferranti for the Manchester electronic computer was 10 inches in diameter and 12 inches long. It stored 16,384 40-bit words in a serial mode of operation. The drum revolved at 2000 rpm, giving a maximum access time of 30 ms. The 40-bit words were used for data, and instructions were 20 bits long. Ferranti later used the same type of magnetic drum in the Mercury computer.

MAGNETIC DRUM DEVELOPMENTS AT IBM

In 1949 an experimental magnetic drum was constructed by tightly wrapping an aluminum drum with a single layer of wire made of CuNiFe, an alloy known to be suitable for magnetic wire recording. The wire layer was machined down to half its original thickness, and then worked smooth. Intended for the initial development of a drum-augmented IBM 604 calculator, this drum was to provide a table lookup capacity of 10,000 decimal digits.

At the same time, James W. Birkenstock of IBM contacted ERA regarding a drum design for a computer. An ERA-IBM agreement was signed on March 8, 1950. ERA furnished the design for a magnetic drum to be used in a so-called Magnetic Drum Computer. An engineering model of such a drum was belt-driven to rotate at 4280 rpm and had a CuNiFe wire surface. This development was followed, in the 1952–1953 period, by the fabrication of preproduction models of drums for the IBM 650 computer. These drums were 4 inches in diameter and 16 inches long and had a plated magnetic recording surface. The drums rotated at 12,500 rpm, providing a maximum access time of 4.8 ms. The IBM 650 computer employed this plated drum for internal memory. It stored 1000 or 2000 biquinary words of 10 decimal digits, plus sign. Biquinary decimals are expressed in binary form as 0 or 5 plus (0, 1, 2, 3, or 4). This mode of expression provides a means of self-checking for accuracy in performing computation.

IBM also employed magnetic drum storage in its series 700 computers. The IBM 701 drum stored 8192 or 16,384 words each consisting of 35 alphanumeric characters, plus sign. The average access time was 50 ms to the first word of a group of words being addressed. The IBM 702 computer drum stored 60,000 alphanumeric characters with an access time of $(8.12 + 0.04N)$ ms, where N is the number of characters in a group being addressed. If $N = 1$ (i.e., for the first character in a serially addressed group), the access time is 8.16 ms. In 1956 the IBM 704 computer was introduced. Its drum stored 8192 or 16,384 words of 36 bits each. The IBM 709 computer used the same drum storage found in the 704. The IBM 705 computer's drum storage of 60,000 alphanumeric characters was similar to that of the IBM 702 machine, storing 60,000 alphanumeric characters with an access time of $(8 + 0.04N)$ ms.

A number of custom-designed, one-of-a-kind magnetic drums were built by various organizations for computers during the 1950s. The minimal automatic computer, built by the Digital Computer Group at the California Institute of Technology, used a drum 6 inches in diameter and 8 inches long revolving at 3500 rpm. It stored 8192 32-bits words, and operated in a serial mode with a maximum access time of 18.7 ms. Marchant Research used a magnetic drum for internal computer storage of 4096 words, each of 10 decimal digits. The drum was 6.5 inches in diameter and 10 inches long and revolved at 6000 rpm, giving a maximum access time of 10 ms. Its first use was in April 1953.

The Naval Research Laboratory in Washington built a computer using a drum 12 inches long. It stored 2048 45-bit words and revolved at 3600 rpm. Operating in a parallel mode, it had a maximum access time of 16.7 ms.

A computing machine was built by the RAND corporation in Santa Monica, California. It was first used by the air force in February 1954. It had a magnetic drum for external storage of 2048 words of 40 bits operating in a parallel mode. At a revolution rate of 1200 rpm, the access times were 25 ms maximum and 5 ms minimum.

Northrop Aircraft Company built a "magnetic drum differential analyzer" using a drum 8 inches long and 8 inches in diameter. It stored 264 32-bit words and operated in a serial mode. It revolved at 3600 rpm, and had a 16 ms maximum access time. The first use of this machine was July 1956.

The Central Institute of the Royal Norwegian Council for Scientific and Industrial Research used a small magnetic drum in its computer. The drum, which resolved at 4000 rpm, had a diameter of 2 inches, and a length of 5 inches. The word length was 32 bits, and the maximum access time was 15 ms.

The details are not apparent in the literature regarding the nature of the recording surfaces of the various drums mentioned above, or the design of the reading/writing heads. However, it can be assumed that the recording medium consisted of an iron oxide or a plated ferromagnetic layer. The early heads probably were made with permeable metallic core yokes, but the later ones used ferrite material for the head cores.

DRUMS FOR PERIPHERAL STORAGE

Sperry UNIVAC produced a large-capacity magnetic drum system called the Fastrand at its plant in Blue Bell, Pennsylvania. J. P Eckert, who, with John Mauchly, had formed the Eckert-Mauchly Corporation, was in charge of the Fastrand development. The storage unit consisted of two plated drums, each about 5 feet long and 18 inches in diameter and mounted horizontally, one parallel to and above the other. The read/write heads for these two drums were mounted on a track that lay between the two drums and parallel to their axes. There were 32 read/write head carriers, each carrier supporting a pair of heads, one head to write or read on each of the drums. The heads were pivoted so that they flew on the boundary layer of air attached to the rotating drums. Thus there were a total of 64 heads to read or write data. The head carriers were moved along the track by voice coils, so that each head serviced 96 tracks. There were 96×64, or 6144 data tracks storing more than 300 million bits. The Fastrand heads were selected by a decoder system. Timing tracks in the middle portion of each drum provide binary signals for controlling the read/write operations. The system was rather sensitive to floor vibrations.

A system of single-drum units employing groups of fixed, flying head blocks was developed at the Sperry UNIVAC division in Roseville, Minnesota. The largest of these, designated the FH-880 is shown in Figure 16–6. The drum

Figure 16–6   An FH-880 magnetic storage drum unit. [*Courtesy of The Babbage Institute and Lockheed-Martin Corporation.*]

was 30 inches long and 24 inches in diameter. It contained 880 recording tracks, which were written and read by 40 head blocks. Each block contained 22 separate read/write heads. Five head blocks were mounted in fixed positions on each of eight rods. There was a pulley attached to one end of each rod and a cord connecting all the pulleys, as shown in Figure 16–6. The pulleys could be turned so that the heads were all elevated from the drum surface if the drum was to be moved or, when in use, before the drum began to rotate. The 22 heads in each block were pivoted, so that when the drum came up to full rotational speed, the pulleys were turned in a way that allowed the heads to nearly touch the drum. In fact, the heads flew on the laminar-flow boundary layer of air at a displacement of 0.0005 inch from the drum. Each of the 880 tracks stored 39,064 bits on a 24-inch-diameter drum coated with gamma ferric oxide. The total data capacity of one drum was over 35 million bits. The rotational speed, 1750 rpm, was such that the average access time for any 22-bit word was 17 ms, and the maximum was 34 ms.

This drum system was released for delivery in 1963. The drum was driven by a motor that was integral with the drum shaft. The system was especially developed to provide external storage for large mainframe computers such as the UNIVAC 1107 and 490. It could serve for mass storage using a single unit or a group of many units.

Other flying-head drum units supplied less storage per drum. For example, the FH-400 had a drum length and diameter of 17 inches, and rotated at 1740 rpm. This drum had 400 tracks, of which 396 held data, and it had a written timing track. There were 18 head blocks, each containing 22 flying heads.

The still smaller FH-220 unit had a 3.3 million bit capacity with stored data on 200 recording tracks. This drum was 9 inches long with a diameter of 10.5 inches, and it revolved about a vertical axis. Information was stored at a bit density of 500 b/in., and the data were written and read by 10 head blocks per drum, each block contained 22 flying heads. The drum turned at 3550 rpm, giving a maximum access time of 17 ms. The FH 500 drum was the same size as the FH-400 but had 24 flying-head blocks, each containing 22 flying heads, for a drum with 528 tracks. This drum also revolved at 3550 rpm, giving a maximum access time of 17 ms.

All these smaller flying-head drums had the same construction for mounting and turning the head blocks and connecting the motor to the drum shaft, as was done with the FH-880 drum unit. The 22 read/write heads in the head blocks were made of ferrite and contained silver or copper foil in the active gap of their core. The core winding consisted of 40 turns of 40-gauge copper wire.

Various newer disk memories, with much higher bit densities yielding more compact, lower cost, very large capacity external storage, have replaced drums for external (peripheral) use. The development of higher density integrated semiconductor devices (chips) provided cheaper access, more rapid by orders of magnitude, and larger storage capacity than could ever be attained with drums for internal memory storage. Drums served a very useful purpose as computer stores before large-capacity semiconductor memories became available. At the time of their early development, there were no other means for providing rapid access to the amount of information required for scientific computation or processing.

REFERENCES

Allen, Alden, Papers, 1950–1971. Charles Babbage Institute No. 26. Specifications, photographs, and service manuals on ERA and UNIVAC drums.

Engineering Research Associates, "Summary Report: Magnetic Recording of Pulses for the Storage of Digital Information," Charles Babbage Institute No. 27, June 19, 1947.

Kilburn, Tom (University of Manchester), interview by C. R. Evans, OH 024. Transcript available at the Charles Babbage Institute, University of Minnesota, 1976.

Pugh, Emerson W., *Memories That Shaped an Industry*, MIT Press, Cambridge, MA, 1984.

Pugh, Emerson W., L. R. Johnson, and John H. Palmer, *IBM's 360 and Early 370 Systems,* MIT Press, Cambridge, MA, 1991.

U.S. Department of the Navy, Office of Naval Research, "A Survey of Automatic Digital Computers: 1953." Contains specifications of several drum memories. Available from the Charles Babbage Institute, University of Minnesota.

Westbee, Robert L., interview by Arthur L. Norberg, audio tape(s) and transcript, North Oaks, MN, OH 112, June 5, 1986.

17 Data Storage on Tape

William B. Phillips

In 1949, everything was in place for the development of digital data recording. The incentive was there in that new electronic calculators were starving for more storage and higher data rates, and a potential solution was taking shape in the form of magnetic tape recording. Tape recording technology for professional audio was showing great promise, following the development of Magnetophon-type equipment by Ampex, and BASF-type coated tapes by the 3M Company and others. High-quality audio tape recorders had almost everything needed for the first digital tape recorder. The tape, reels, heads, and transport mechanisms, including drive motors and reel motors, were all appropriate.

The obvious limitation of the audio recorders was that the data rates and access times were much too slow to feed the new high-speed electronic computers. Early vacuum tube electronics and relays had the capability to do thousands of additions per second and dozens of multiplications or divisions per second. Unfortunately, the available storage for intermediate calculations using punched cards or paper tape was only a few hundred digits, and the input/output speeds of these mechanical storage devices was only about 100 characters per second. To feed high-speed electronic systems called for data rates in the thousands of characters per second. Early audio tape recorders represented an improvement, but their digital capability was limited to about 200 characters per second because of low tape speed and single-track recording. Audio recorders were just a starting point, and special magnetic tape systems had to be designed and developed to meet the specific needs of digital data recording.

In early electronic calculators and computers, tape storage became the second level in the storage hierarchy. It served as an extended memory in the absence of any other viable storage technologies. With the emergence of lower cost ferrite core memory, and then disk storage systems, magnetic tape became a fast input/output, archive, and journaling storage technology. Subsequently, with the development of semiconductor memory and large inexpensive disk capacity, the role of tape narrowed even further to archiving plus dump (backup) and restore applications. Finally, with the emergence of desktop computing systems and the personal computer, the size and cost of tape storage had to shrink accordingly, and the application remaining for most systems was dump and restore. Nonetheless, tape remains a storage technology of choice in that it is the lowest cost option and has excellent input/output data rates. When combined with library and storage management software, tape storage is a valued component of the best storage hierarchies across systems of all types, small and large.

FIVE DECADES OF HALF-INCH TAPE STORAGE

The history of digital magnetic tape recording started with 0.5-inch tape. The initial choice of the type of tape medium was between a metal base and a plastic base. Tape was a natural extension of the wire recorder, allowing better recording efficiency through the use of parallel recording tracks and a better packing efficiency. Plastic-based tape offered lighter weight and lower cost, but it was originally thought to be more fragile than metal-based tape.

The first digital tape recorder was the Uniservo I, developed in the late 1940s by the Eckert-Mauchly division of the Remington Rand Corporation. This first product used 0.5-inch metal tape, recorded at 128 b/in., with a 2.4-inch inter-record gap and a data rate of 12,000 characters per second (12 kc/s). The product was announced in 1950 and shipped in 1951. The metal tape was removable from the drive and had a maximum capacity of a few megabytes (MB), depending on fixed-block architecture record size.

It is interesting to compare this early 0.5-inch magnetic tape system with what is available today. A modern 0.5-inch cartridge tape transport, with integrated controller, has a capacity of 10 to 50 gigabytes (GB) per cartridge, depending on the recording system. The integrated drives and controllers are about one-tenth the size of the original tape systems and have a volumetric storage density four to five orders of magnitude greater than that of the original Uniservo.

This chapter describes the evolution of early products and traces how tape recording has maintained a place in today's storage hierarchy. We will also look at formats other than 0.5-inch tape that have emerged as the data processing industry expanded from mainframes to include today's servers, high-speed workstations, and desktop computing systems. Each of the new tape formats came in response to various market conditions that created opportunities for removable storage. Tape remains a basic storage element for removable data storage. It has had a long and prosperous history, with many innovative components, architectures, and products. The state of the art features removable cartridges with capacities up to 50 GB, and data rates of at least 10 MB/s on tape drives that have volumes of a fraction of a cubic foot. The storage device costs are lower than the original pioneering systems from Remington Rand and IBM.

THE UNISERVO I: THE FIRST DIGITAL TAPE RECORDER

The Uniservo I was the first digital magnetic tape recorder designed, in 1951, to be part of a computer system. It was originally designed by the Eckert-Mauchly Company, which was subsequently bought by Remington Rand. The larger company integrated the tape drive into the Remington Rand UNIVAC computer and built a system for input/output for the UNIVAC series. Input was from card or key entry to tape on the Unityper. Then the tape was moved to the Uniservo I tape drive for reading into the tape control section of the UNIVAC at roughly 12 kc/s. Output was written to tape on the Uniservo I, and the tape was subsequently moved to a tape-to-printer system, the Uniprinter. This means of buffering the slow input/output rates of punched cards, key entry, or printers was typical of large data processing systems through the early 1970s. Later, the lowered cost of solid state buffer memories allowed consolidation of these processes into the main-frame computer system.

The Uniservo I was the basis for tape drive development over the next three decades. The basic architecture featured a removable reel of tape, which was mounted and then threaded to a reel on the tape drive. The tape was automatically loaded into the mechanical system for tensioning and moving up to operating velocity across a magnetic head assembly. The operations of writing, reading (both forward and backward), backspacing, spacing/searching forward, and rewind/unload were essentially the same as the operations of tape subsystems today.

In a 1952 paper, Welsh and Lukoff of Remington Rand stated that "a practicable method of obtaining input-output speeds for digital computing devices is the high-speed tape recording method, but the designing of a good tape system has been, to say the least, extremely difficult." They went on to describe the design of the Uniservo I, which had the following characteristics: a density of 128 b/in., an 8-track parallel head, 100 in./s velocity, 10 ms start time (from zero to full velocity), and 10 ms stop time. The final result was a data rate in and out of the UNIVAC computer of 12,800 6-digit characters per second. The mechanical design used a center drive capstan with variable loops of tape on each side controlled by pulleys and arms (Fig. 17–1). The arms controlled the tension of the tape and the file and machine reels during acceleration and steady running. The system used metal tape rather than the plastic-based tapes then available in early audio systems. Metal tape was more robust than the early acetate-backed tapes, but had its own problems of excessive weight, finger cuts, and tape head wear. To

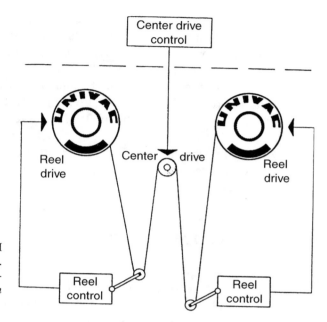

Figure 17–1 Schematic of the Uniservo I tape path, with tension and reel controls. The reel control pulleys were spring-loaded to the proper tension. [*From Welsh and Lukoff (1952).*]

cut down on tape head wear, the Uniservo I used a plastic tape that ran very slowly between the head and the metal tape!

Other architectural aspects included reading backward, overlapped read and write capability (with two tape drives), fixed blocks (720 characters), certified tape with punched holes to identify bad areas, a "sprocket track" to allow for variable density, common read and write circuits for all tape drives, and a small mercury delay line to help buffer the tape data rate to the system. In total, the Uniservo architecture was a very solid, if difficult, design that would not change much until the introduction of large-cache buffers in the late 1970s.

IBM ENTERS THE COMPUTER MARKET

As the UNIVAC system emerged, IBM, which had been tracking the progress of ENIAC and EDVAC at the Moore School of Engineering, was preparing to enter the electronic computer market. Eckert and Mauchly left the University to start their own company, Eckert-Mauchly Computer Corporation, and build their first product, the UNIVAC. The Remington Rand Company then acquired Eckert-Mauchly and established the UNIVAC division of Remington Rand. At that time, IBM was heavily engaged in punched-card calculators; the company had a good customer base and a feeling for the weaknesses of the punched-card architecture. IBM's engineers, too, recognized the need for storage devices that could work at data rates approaching the system data transfer and processing rate. In due course,

IBM designs for mainframe-class computer systems became the de facto standards, and its input/output systems also became standards. As a result, the IBM design choices for these early systems lasted for over three decades.

But returning to the 1940s, the storage medium of the time was the punched card. The punched card held 80 alphanumeric characters and was read at a typical rate of 100 cards per minute, giving an effective data rate of 133 c/s. IBM's entry into tape storage was the model 726 drive with a data rate of 7500 c/s—over 50 times that of cards. Also a single reel of tape could hold the equivalent data of 18,000 punched cards. This combination of speed and compactness made it obvious that the next generation of computer systems should try to exploit magnetic tape technology.

THE IBM 726 TAPE DRIVE

In the late 1940s, IBM had little knowledge of magnetic recording and started with a blank sheet of paper in the design of its first product. The company's experience in punched cards and associated calculators gave its executives a feel for the commercial problems, and influenced the choice of an architectural design. As a new, removable storage medium, the magnetic tape had to offer cost, capacity, and data rates significantly better than those of punched-card systems.

The first IBM magnetic tape system had numerous innovations and was competitive with punched cards in cost, but it was far superior, by orders of magnitude, in capacity and data rate. Innovations included vacuum column control of tape motion and tension, multitrack parallel heads, NRZI encoding format, peak detection, multicapstan/pinch-roller drive, and a non-fixed-block recording format. The recording medium was coated plastic tape, and a tape leader was used for ease of threading the tape into the transport. The tape chosen was the cellulose acetate based, iron oxide coated tape that was commercially available in the early 1950s for the growing professional audio business. The original design used 0.25-inch tape with a 2-track serial recording architecture (Fig. 17–2). To achieve high linear density, a modified non-return-to-zero recording code was chosen, named NRZI. A 7-track parallel format on 0.5-inch tape improved the data rate.

The mechanical inertia of plastic-based tape was much lower than that of metal tape. This translated into shorter vacuum columns, shorter record gaps, which would give better recording efficiencies, and lower power reel motors. The use of plastic tape meant an overall simpler mechanical transport. Early on, the key requirement of the tape transport design was to handle tape gently as it was accelerated from standstill up to the operating velocity, typically 75 to 100 in./s. In addition, it was necessary to maintain a constant tension in the tape to form a well-defined tape–head interface. The tension had to be optimized to give the best

Figure 17–2   An early prototype (October 1951) of the IBM 726 with quarter-inch tape, vacuum columns, and a vacuum motor in front of the tape drives. [*Courtesy of IBM Corporation.*]

trade-off between tape wear, head wear, and minimum magnetic separation at the interface. The recording engineer would like the tape pushed into contact with the head to give the best recording performance, whereas the mechanical engineer would like the tape removed from the head to extend head and tape life and to reduce friction. Practical designs were a continual compromise between these two sets of requirements.

With the high accelerations required to get tape up to speed from a dead stop, the vacuum column tape control system was developed by Jim Weidenhammer in the early 1950s and patented in 1962 (Fig. 17–3). It was a unique solution that provided essentially uniform tensions across the tape and allowed the tape to be brought up to full speed in 7.5 ms on the first two systems, and around 1 ms in later ones. The gentle handling enabled the use of the more fragile, but lighter, plastic-backed tape. This was an unbeatable combination for high-performance tape systems and was used from the 1950s through the 1970s. The vacuum

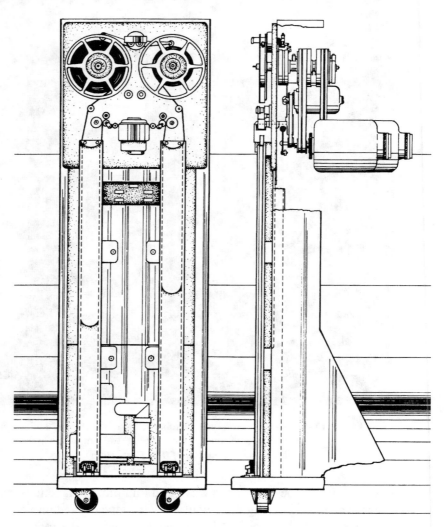

Figure 17–3   Tape feed mechanism patented by James Weidenhammer and Walter Buslik showing the vacuum columns and the basic tape path schematic of the 726. The patent was filed in May 1952 and issued in October 1962. [*Courtesy of IBM Corporation.*]

columns had sensors along their length that allowed full control of the reel motors. The length of the columns had to allow for the worst-case conditions, where tape could be almost instantaneously reversed in a backspace operation. To contain the worst-case start, stop, and reverse velocity conditions, for velocities in the 100 in./s region, the vacuum column was several feet long.

The choice of 7 parallel tracks on the tape required a head capable of recording and reading back all 7 tracks simultaneously. The original design used one head per track to write the data and subsequently read it back. There was already a precedent for a multitrack head in the Uniservo, albeit those tracks were staggered for ease and accuracy of manufacture. At IBM, processes were developed to manufacture 7 in-line tracks across the recording head.

With the vacuum columns providing the tension control, the basic drive mechanism was a symmetrical pair of drive capstans providing a tape velocity of 75 in./s, with the recording head placed approximately midway between the capstans. The tape was pinched between one or the other capstan and an idler driven by a moving coil actuator, biased to the forward or backward direction. The moving coil, magnet, and magnetic circuit were designed to move the idler mechanism into either a drive, neutral, or stop position as fast as possible. The faster the tape could reach terminal velocity, the better the access time and the smaller the inter-record gap.

Following the Uniservo design, as much as possible of the common electronics and power went into the control unit, to minimize cost and space. The electronics of the early 1950s was based on vacuum tubes, along with relays, to perform all the logical functions of the tape systems. Later, crystal diodes were used to provide more compact and energy-efficient logic electronics. Nonetheless, the early tape systems used hundreds of vacuum tubes in pluggable units of eight tubes.

Early on in the recording system development, all the encoding techniques for digital data had problems. Either there were more transitions per bit of data than needed, lowering the upper limit of the recording data density achievable and reducing the recording margin to perturbations, or there were too few transitions to provide reliable clocking data to keep all 7 tracks in synchronism. Byron Phelps was the project leader of the recording engineering team, and he developed a new code he named NRZI, a form of non-return-to-zero encoding. This code is still used today because it provides a good compromise between the transitions per bit and the data clocking needs.

The NRZI code is not self-clocking and requires the recording format to provide some means of clocking data. Thus a complementary track was recorded in the second track of a 2-track parallel system. The second track was recorded with exactly the opposite signal to the data track, so that a "one" was guaranteed in one or the other track, providing the necessary clock bit per character. This limited the data rate to 7500 b/s, or 1250 (6-bit) c/s. While fast compared to the punched card, this data rate was well below the interface speed and the computer data rate.

The 2-track system was soon replaced with a 0.5-inch tape drive with 7 tracks in parallel: 6 data tracks plus an odd parity track. The choice of odd parity gave a minimum of one bit per character, satisfying the need for clocking data. The latter was true only if all the bits in a track were read prior to any bits of the subsequent character being read. When bits from other characters were read early, it was mainly because the guiding accuracy of the tape path was inadequate. If the tape "fishtailed" through the head guide system, the linear bit density achievable was reduced.

IMPROVEMENTS IN HALF-INCH TAPE SYSTEMS

The original IBM 726 and 727 systems using 10.5-inch reels and 0.5-inch tape became de facto standards for commercial and scientific computer systems. Numerous other tape systems emerged, but they used the standard format of these early systems to facilitate tape interchange. Systems manufacturers, including National Cash Register and Control Data Corporation, had tape systems using this format. In addition, equipment manufacturers, including Ampex, Potter, and Mohawk Data Systems, offered tape systems using this format. Other formats were developed, such as the Sperry Rand Uniservo, which had innovative features, but they never became a standard in the industry.

IBM TAPE DRIVES USING 10.5-INCH REELS

The operating characteristics of early IBM tape drives, using 0.5-inch tape and 10.5-inch reels, are listed in Table 17–1.

These product developments were all driven by data rate, and the easiest way to increase data rate was to increase the linear density. This also increased the capacity of a reel of tape, in some instances far beyond the customers' capa-

Table 17-1 Operating Characteristics of the Early IBM Half-Inch Tape Drives

CHARACTERISTIC	TAPE SYSTEM (Year)						
	726 (1953)	727 (1955)	729-3 (1958)	729-6 (1962)	2401-6 (1966)	2420-7 (1969)	3420-8 (1973)
Velocity, in/s	75	75	113	113	113	200	200
Density, b/in	100	200	556	800	1600	1600	6250
Data rate, kB/s	8	15	63	90	180	320	1250
Capacity, MB/reel	3	6	16	23	46	46	180
Interblock gap, in.	0.75	0.75	0.75	0.75	0.6	0.6	0.3

bility to use it. A "cost performance " measure was used equal to the tape system cost divided by the data rate. As can be seen in Table 17–1, the development of data tape recording technology over 20 years produced a storage capacity gain of over 60 times and a data rate increase of over 150 times. The special needs of data storage reliability resulted in a number of technology innovations to improve the timing accuracy of signals from parallel tracks and the amplitude fidelity of all the recorded tracks. These innovations led to vastly improved data coding methods and error correction techniques. Hardware improvements, such as dual heads to verify the recording accuracy, and electronic deskewing to improve the timing of signals from parallel tracks, were invoked during this period of tape drive development. Finally, the tape velocity was substantially increased, and the start/stop times were greatly decreased to improve the data rates and access times.

NEW FORM FACTORS FOR DATA TAPE STORAGE

During the 1980s, semiconductors achieved the cost and speed that allowed trade-offs between electronic and electromechanical parts of the system. In addition, the microprocessor ushered in the era of the personal computer as well as driving new product design into the software logic era. The personal computer was so price sensitive that cost became the dominant criterion in all storage systems. In addition, the storage hardware had to fit the available form factor in these desktop systems. The slots available for storage devices started with the 5.25-inch-diameter disk drives, and went down in size from there in 70% reduction steps to the current 1.8-inch form factor. Now not only cost but size would be a driving factor in these systems.

THE IBM 3480 TAPE CARTRIDGE FAMILY

IBM introduced a new concept for 0.5-inch tape storage with the 3480 system, the first of a family of products that used a new small tape cartridge in place of the venerable 10.5-inch reel. Computer systems were shrinking in size, and the personal computer was growing in popularity. Big data processing customers typically had large rooms near or on their raised floor areas to house tape libraries consisting of thousands of 10.5-inch-diameter tape reels. It was not unusual for a customer's library storage to have 20,000 to 100,000 reels of tape.

The 3480 cartridge was designed to reduce the storage space per reel or cartridge, and to exploit available media technology used in the consumer video recorder world. The cartridge was roughly $4 \times 5 \times 1$ inches in size, less than

25% of the volume of a 10.5-inch reel (Fig. 17–4). It would barely fit in the new 5.25-inch form factor then becoming popular for hard disk drives. The cartridge used a new, high-coercivity, chromium dioxide tape. Chromium dioxide was developed by Du Pont in the late 1960s and was licensed to Memorex, BASF, and Sony, with commercial audio and video tapes being marketed in 1970. This tape was chosen for the IBM 3480 in the 1970s, when the competing doped gamma ferric oxides were having magnetic stability problems. Chromium dioxide had a reputation for high head wear, and IBM worked hard to dispel that image of its chosen tape. The particle was in fact more abrasive than the old gamma ferric oxide but, with an improved binder, surface finish, and lubrication system, the wear rate of the new medium was brought under control.

Figure 17–4   A 10.5-inch reel, the 3480, and 3480E cartridges. The 10.5-inch reel typically had 2400 feet of tape and, at a density of 6,250 bytes per inch, had a typical capacity of 150 MB. The 3480 cartridge held 540 feet of tape and had a typical capacity of 200 MB. The 3480E had 1100 feet of tape and a typical capacity of 400 MB. [*Courtesy of Imation.*]

The tape drive system had a relatively slow start time of over 10 ms, but the slow access time was masked by use of cache buffers. As electronic speed and memory capacities increased, it had become feasible to buffer the system to allow the tape to come up to speed while data in the buffers were being read out at the channel rate of 3.0 MB/s. This allowed the elimination of the power-hungry large reel motors, the long vacuum column, and related control systems. The result was a compact, low-cost, yet high-performance tape transport.

The recording format featured an 18-track head that could record 2 bytes wide across the tape width. The readback element used a flux-sensing, magneto-resistive (MR) element, so that the signal output was independent of tape speed. The 3480 was originally designed for a tape speed of 40 in./s, and the MR head was therefore essential to get sufficient output at the bandwidths involved. Later, the system velocity was doubled to 80 in./s to operate at the new channel and disk drive rates of 3 MB/s. The MR head technology still had a marginally better signal-to-noise ratio than a conventional inductive read head. The use of the MR technology was a first for digital tape systems and represented the first major successful application of a magnetoresistive read head in magnetic recording.

The 2-byte wide parallel recording was part of an encoding and error correction system dubbed "adaptive cross parity." This was a group coding format that buffered over a hundred bits to give essentially 2-track error correction and 3-track error detection. The resulting reliability at the 19 kb/in. recording density was over 10 times better than for the prior technology, providing about a terabyte (10^{12} bytes) of data per hard failure.

The 3480 tape drive was a landmark in the long evolution of magnetic tape drive cartridges for data storage. It represented the first major departure from the 35-year old reel-to-reel recorder by introducing the single reel cartridge. In addition, it ushered in the magnetoresistive head, which is well on the way to displacing inductive read heads in all data tape and disk storage devices.

CONSUMER RECORDER APPLICATIONS

Consumer recording has provided the technology base for numerous data tape storage products starting with the Uniservo I. Early on, the Philips audio cassette was the basis for several products that focused on replacing the punched card as the input to data processing systems. Later, with the emergence of the personal computer and the move of mainframe computing power to the desktop, other consumer recording and electronics technologies were adapted for new data recording capabilities, including digital audio tape (DAT), 8 mm camcorders, and CD-ROMs. Key to the acceptance of the resulting lower performance data recorders was the ability of the cartridge to back up or store the entire capacity of the local disk drive on the desktop. Performance measures changed from kilobits per second to gigabytes per hour. The latter give an easy measure of the capability to dump a local disk onto the tape cartridge in the allowable backup window. For personal computers, it can take an entire evening to fill a high-capacity, low-data-rate cartridge. For example, a typical 2-GB disk can be dumped to tape with a data rate of 250 kB/s in just over 2 hours. However, as more powerful desktop systems started using these low-cost systems, there was a need for better performance. The disk capacity was increasing and, at the same time, the available system backup time was decreasing.

The Philips compact cassette, introduced in the 1960s, was the first cassette design in which the head was inserted into the cassette for reading and writing, and a capstan was inserted and pinched the tape against a roller to drive the tape. The reels were mechanically coupled to drive motors in the drive to provide the necessary tape tension for the read/write process and fast forward/backward tape spooling. The system was, and is, a consumer success, selling millions of tape drives and cassettes. There were numerous attempts to use the cassette in various tape transports to give a low-cost, medium-performance data recording system. The data cassette, developed in the mid-1970s, formed the basis for low-capacity, low-data-rate, low-cost data recorders. While there are still digital data storage systems that use the cassette, the low performance and the lack of cassette robustness precluded it from becoming a standard in any of the high-performance computer systems.

The Ampex Corporation used its professional video recorder as a base for the Terabit Memory (TBM) system announced in 1969. The company was able to exploit the high recording density, data rate, and capacity of its professional video system using an array of up to 32 helical-scan tape transports. The system had a capacity of 90 Gb to 3 Tb of data. The system had wide appeal to the scientific community, with its unquenchable thirst for data bandwidth, and many of the federal laboratories and scientific computer centers used these systems. However, the Terabit file had little commercial acceptance among computer centers. As described in an earlier chapter, the technology came from the Ampex Quadruplex video recorder and used 2-inch tape with a transverse-rotating head. Data were recorded twice for reliability and had a 750 kb/s data rate. Each transport had a tape with a capacity of 90 Gb (roughly 10 GB). The system could be configured with as many as 32 drives, giving a total capacity of 3 Tb. The data were recorded in large sequential bursts, and the TBM's format was perfect for scientific data storage.

In the mid-1980s, Honeywell developed a helical-scan data recorder using the popular VHS television cassette. The capacity of a VHS cassette was several gigabytes, and the data rate was in the 500 kB/s range. The system's performance specifications were excellent, but the VHS cassette did not appear to be robust enough to meet many customer environments. The typical commercial and scientific customer had been sold on tough, robust cartridge designs and outstanding data integrity.

In the late 1980s, the digital audio tape system (DAT) was developed using a helical-scan transport and a relatively small, 8 mm tape cassette (0.4 × 2.9 × 2.1 in.). The first DAT audio recorders were introduced in late 1986 in Japan and later in Europe. This unique consumer product was developed to replace, ultimately, the analog Philips cassette. But the DAT technology had another competitor in the audio market in addition to the ubiquitous Philips cassette, and that was the emerging compact disc. The latter used optical digital recording to give outstanding audio performance. Although the DAT has met with limited commercial success in the audio market, the data storage spin-off has enjoyed good acceptance in the personal computer market, starting in 1988. Early on, Hewlett-Packard and Sony had collaborated on a DAT-based format they named digital data storage (DDS). They introduced a series of drives over several years that had capacities from 1 GB per cassette to over 10 GB per cassette. At one time in the early 1990s there were up to 20 manufacturers of DAT data recorders. In the last several years, company failures and consolidations have reduced that number to 10. However the unit volume of tape drive shipments remains healthy and continues to grow. Several other recording formats were developed using the basic DAT cassette, but the DDS format enjoys a large market share today.

In the late 1980s, Exabyte developed a new tape transport based on the 8 mm video cassette camcorders available from numerous manufacturers, modified for data recording. The initial product, introduced in 1987, had an excellent capacity of 2.3 GB per cassette and a low, but acceptable, data rate of 250 kB/s. Contrary to the industry consensus, the 8 mm business grew and prospered under the leadership of Exabyte. Exabyte has enhanced the product over the past decade, and the 1997 Mammoth has a capacity of 20 GB per cassette and a transfer rate of 3 MB/s.

LOW-COST SOLUTIONS

The 3M Company developed a unique 0.25-inch cartridge (QIC) in the mid-1970s. This was 3M's answer to the low-performance problem of the data cartridge, based on the Philips cassette. Like the Philips cassette drive, a head and capstan were inserted to drive the tape, but tension was controlled internally by a viscoelastic belt inside the cartridge. The belt provided good tape tension and squeezed air out of the reel stacks, allowing for high-speed tape motion in the vicinity of 100 in./s. The cartridge had internal guides to provide good registration for multitrack recording, and these guides, and the reel system, were mounted to a rigid metal plate that became the reference for the drive mechanism. Having the tension and guiding mechanism internal to the cassette allowed for significant tape drive simplification and cost reduction. The 3M Company developed the cartridge and manufactured drives for the OEM market in 1981. Very quickly, a manufacturers' group was formed to deal with interchange standards for the new format. The drive and cartridge were ideal for the emerging personal computer market, particularly since the cassette could be removed without rewinding, and reloaded into the drive in essentially the same place it left off, precluding a search operation. The QIC format met most of its expectations and has been very competitive since its introduction.

In 1985 the Digital Equipment Corporation (DEC) developed a low-cost, 0.5-inch tape drive, the TK50, which used a single-track head indexing between 22 positions across the tape. The head was indexed to the proper track to give a serpentine recording format, and the single-reel cartridge used a leader for threading and loading into the tape drive. The 45 kB/s data rate was at the low end of the performance needed for a backup tape system. The drive was originally designed specifically for DEC systems but, later, a series of products was introduced that featured improved capacity and data rate, culminating in the introduction of the digital linear tape (DLT) drive. The DLT was intended mainly for DEC systems, but in 1994, DEC sold the DLT and other storage products to Quantum Corporation. Since that time the DLT technology has enjoyed rapid growth in the small and medium systems backup applications.

TAPE LIBRARIES AND AUTOMATION

Starting with the first data storage tape systems, automation was an obvious enhancement. For example, in the late 1950s, IBM shipped its Tractor tape and library system to the U.S. government. Although Tractor was technically successful, it was decades before any tape library would prove commercially successful. In the early 1970s, the XYTEC 0.5-inch tape library stored 10.5-inch reels and used robotics to deliver them to conventional tape drives. A later product, the IBM 3850 with its unique cartridge, helical-scan drive, and disk staging architecture had some success. But both the XYTEC and the 3850 had short product lives, and no follow-on products were made.

In the mid-1960s, there was a vision of a magnetic tape library that could be totally automated and controlled by the mainframe. Earlier attempts at tape libraries either had poor commercial acceptance or did not seem to be what the typical commercial customer needed for a storage environment. The IBM Digital Cypress system had a tremendous following in the scientific world but not in the commercial arena. An architecture emerged that featured the size of the Cypress cartridge, and the entire system included a library, several cartridge transports, and suitable controllers. After many design changes, the IBM 3850 Mass Store emerged in 1975, featuring a small cartridge, a library, a helical-scan transport, and a new architecture in which the 3850 served as a disk staging storage system. All data recorded and read back on the 3850 cartridge system had to go through the mainframe disk system. This gave the system the look and feel of a storage hierarchy in that the mainframe main memory, the disk subsystem, and the 3850 cartridge subsystem were all controlled as a storage hierarchy. The controlling software used various storage algorithms to determine which data sets would migrate to the various levels of the system for optimum performance. The ultimate goal was to get disk-like performance at costs approaching those of magnetic tape. The 3850 cartridge had 800 feet of 2.7-inch-wide tape, wound into a conical cartridge with a rounded "nose." The shape was designed to give reliable insertion into the storage cells of the library. The capacity was 50 MB, the capacity of an IBM 3330 disk drive. But, by the time the 3850 hit the market, the next generation of disk drives was already in production, with a capacity of 100 MB per drive. Unfortunately it then took two 3850 cartridges to "dump" a disk drive.

The library was a long linear array of honeycomb-like cells that held hundreds of the 50 MB cartridges. Access was by a pair of robotic units that picked a cartridge from the cell and delivered it to one of several helical-scan transports mounted on the library walls. Each robot had an X-Y mechanism and sensors to access any of the cells on either side of the library. The system could get very close to disk-like performance if there were good data set "miss ratios" at the main memory, cache, and disk levels of the system. The 3850 was a pioneer in hierarchical data storage management, but had limited commercial success. It did not achieve broad customer acceptance, and IBM offered few enhancements to the basic hardware and software. It was reported that as many as 700 systems were shipped, but as many as half of these were used within IBM.

In 1987 Storage Technology Corporation (STK) introduced its version of the IBM 3480, together with a unique cylindrical library (called a silo by most customers). Storage Technology had just emerged from bankruptcy and gambled the company on the new library. Customers had to use STK (not IBM) drives if they wanted the library and related management software, and over time the library was attached to over 20 non-IBM mainframes through a serverlike architecture. One of the benefits of automation was that customers were willing to archive their old data previously stored on disk, and they found that automation gave access to their tape records in less than a minute. With manual archives it was more likely hours or days before old archives could be put online. The result was a freeing up of disk space as old data was moved to tape libraries. Storage Technology's system included extensive software for managing the customers' data sets in their libraries. Over time and several library enhancements, this data management software has grown extensively. This was a preview of the emerging importance of data management software for the storage systems of the 1990s.

OTHER IMPACTS ON TAPE STORAGE SYSTEM DESIGN

A number of hardware and software innovations have changed the way tape storage systems are designed. Helping to make this possible were the continued improvements in semiconductor technology, both in high-speed circuits and in DRAM memory. Both circuits and memory are the basis for the data caching architectures now prevalent in data storage devices. Caching gives the performance of memory at the cost of disk or tape storage. By the clever use of memory, tape drives, essentially sequential devices, can use lower performance mechanics to give smaller, lower cost, and more reliable systems. In addition to caching, powerful data encoding, compression, and correction algorithms are now viable as a result of the speed and low cost of electronics. Interfaces now batch the input/output streams to move data on the electrical interface at maximum speed.

With the growth of personal computer power, new cost pressures grew on all aspects of the system. Price points had to be maintained to compete for the consumer dollar. As a result, products targeted for these markets had to maintain or reduce their price, and large volumes were required to meet the price targets. Tape systems are rare in the home environment, but common in offices. The cost of lost or unretrievable data to a company is worth the investment in an inexpensive dump/restore device. Performance, as such, was not a high priority provided the local disk capacity could be dumped to tape in 8 hours or less. Disk costs per megabyte dropped, and disk capacity grew dramatically. In 1990 hard disk capacity was commonly 40 to 80 MB; in 1997 disks held 2 to 9 GB. The tape storage required has to be able to contain this type of growth.

In 1991 the concept of moving error correction techniques from the individual data streams back to the basic storage unit was implemented by means of small hard disk drives. The concept was named RAID, for "redundant array of inexpensive disks." The idea was to take advantage of the low cost per bit of the 3.5- and 5.25-inch disk drives and increase the overall system speed and reliability by spreading data across an array. The same techniques have been applied to tape drives and tape libraries (RAIT and RAIL), where data can be spread across several tape cartridges or libraries using an ECC algorithm. Again, the concept is to exploit low-cost storage devices in a redundant array to achieve dramatic improvements in reliability. The redundant array products have shown impressive results in high-performance disk systems, but it remains to be seen whether these concepts will be successful in the cost-sensitive tape market.

The concept of storage hierarchies and the storage management required to take advantage of them was conceived in the late 1960s when processor memory, disk, and tape had to coexist on the typical computer of the day. The same concepts and the same storage technologies are still the top three levels in today's typical storage systems. Again the objective is to achieve the speed of the fastest level of the memory hierarchy at the overall cost of the cheapest level of the hierarchy (tape).

SUMMARY

Magnetic tape storage has been a valued member of the storage family for many years. However, as storage costs drop at each level of the hierarchy, tape systems will be under competitive pressure from disk, as well as from other potential tape replacement technologies, such as optical disk or holographic storage. Tape has to compete in terms of cost, performance, and overall customer satisfaction. The latter category raises the points that tape is not error free and requires some special handling to achieve the best possible reliability. There are two ways to counter these disadvantages: to maintain a one or two orders-of-magnitude cost advantage, or to improve the reliability to exceed that of the competitive technology. From the desktop to the small server systems, tape is used solely for backup. This single application will drive high data rates as backup windows shrink, as long as the cost of the data rate is competitive. At the high end, many of the older tape applications are changing or disappearing as lower cost disks are used more effectively. More optimistically, the incremental cost of tape makes it more than competitive to replace other removable storage such as microfiche, microfilm, and paper. To be viable, this replacement must be combined with automation and data management software. The opportunity here is vast, since it is widely accepted that paper represents up to 95% of the data stored today.

Numerous studies of the storage density capability of magnetic tape and magnetic disk systems have shown we are not near any immediate natural limit imposed by the physics of the recording system. Some practical limitations pose short-term problems, but there appears to be no immediate risk of reaching a dead end with capacity, performance, or reliability.

REFERENCES

Harris, J. P., W. B. Phillips, J. F. Wells, and W. D. Winger, "Innovations in the Design of Magnetic Tape Subsystems," *IBM Journal of Research & Development*, **25**, 690–699 (1981).

Johnson, L. R., J. H. Palmer, and E. W. Pugh, "Strength in Storage Products," in *IBM's 360 and Early 370 Systems,* MIT Press, Cambridge, MA, 1991, pp. 220–240.

Pugh, E. W., "Storage Hierarchies: Gaps, Cliffs, and Trends," *IEEE Transactions on Magnetics,* **MAG-7**, 810–814 (1971).

Welsh, H. F., and H. Lukoff, "The Uniservo-Tape Reader and Recorder," *Proceedings of the Joint AIEE-IEE-ACM Computer Conference*, New York, 1952, pp. 47–53.

18 Data Storage on Hard Magnetic Disks

Louis D. Stevens

In June 1957 IBM shipped a revolutionary new product, called RAMAC (random access method of accounting and control). The system was revolutionary because it contained the first commercial disk drive, the IBM 350. The 350's magnetic recording technology, based on that for magnetic drums, permitted an areal density of 2000 bits/in.^{2} (100 bits/in. and 20 tracks/in.). This level of areal density was possible because the magnetic read/write head was supported above the disk surface by a hydrostatic (pressurized) air bearing.

This chapter contains two sections: the first describes the development of the IBM 350 disk drive, and the second discusses the evolution of subsequent magnetic disk products and related recording technology.

Improvements in air bearings and magnetic recording technology since 1957 have led to magnetic disk products in 1997 that achieve areal densities of 2.64 Gb/in.^{2}, a more than 1.3 million-fold increase over that of the 350. Figure 18–1 plots the areal density increases achieved by the end of each five-year interval from 1957 to 1997. The more than six orders of magnitude increase represents, on the average, a doubling of areal density capability every two years. This increasing areal density capability can be attributed, mainly, to reduced spacing between the magnetic read/write head and the magnetic disk surface; and secondarily to the reduced length and width of the read and write gaps, decreased thickness and increased coercivity of the magnetic disk coating, increased precision in locating and maintaining the read and write gaps over the centerline of the disk data track, and improved signal encoding/decoding techniques. From 1957 to 1991 these factors led to an average annual areal density growth rate of about 39%, but from 1991 to 1997 the annual growth rate increased to about 65% with the help of new magnetoresistive read heads and

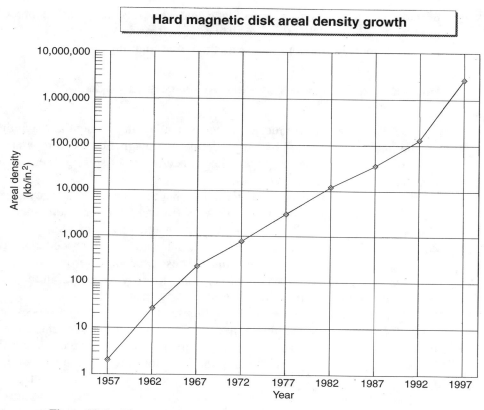

Figure 18–1 The areal density "score card" for magnetic disk products shows the highest areal density achieved by the end of each five-year interval since 1957.

smaller diameter disks, which made practical sputtered-alloy thin-film disks with superior magnetic properties and better physical surfaces; the smoother surfaces, in turn, permitted lower flying heights.

THE DEVELOPMENT OF RAMAC

In this section we describe the start-up of a new laboratory, and its initial search for projects that would enhance IBM's punched-card accounting systems. We trace how one of those projects evolved to the RAMAC system and its 350 disk file. We discuss how the project went through a basic metamorphosis from a normal development effort, directed at improving the cost and performance of an existing function by the application of new technology, to a development directed at providing a new way to process data; and how this new way to process data became a holy grail for the new laboratory. It was even reflected in the name RAMAC, short for random access method of accounting and control.

Processing Data Files Stored on Magnetic Tape

We begin by briefly recalling the sequential file, sequential batch data processing technique adopted by early commercial magnetic tape computer systems to reduce to a tolerable level customer concerns about storing operational business data on erasable magnetic tape.

The first commercial computer to use magnetic tape for storage and processing of large data files was the UNIVAC I (**uni**versal **a**utomatic **c**omputer), first shipped in June 1951. It and subsequent competitive commercial magnetic tape computers used sequential file and batch data processing procedures similar to those used with punched-card machines. These procedures provided backup files at all times, as well as recovery, without data loss, by restarting, in the event of a malfunction on any step of the process.

Let us take a close look at using this sequential batch data processing technique for maintaining inventory files on magnetic tape. A sequential master file on tape contains a record for each part in inventory, with the records in part number sequence. Transactions affecting the inventory are accumulated on a second tape over some period, such as a week, or a month. This transaction tape containing additions, deletions, and changes is sorted so that the records are also in part number sequence. After backup copies of both the old master and transaction tapes have been made, records from the transaction tape are processed in the CPU, record by record, against the master tape records. The part number fields from the transaction record and the master record are compared and one of the following four actions is taken:

1. For a change: update the old master record in the CPU and write it to the new master tape.
2. For a new part: insert a new record on the new master tape.
3. For part deletion: delete the old record in the CPU.
4. For a part not active: copy the old record in the CPU to the new master tape.

Maintaining a sequential file in this manner produced a new updated master tape, leaving the old master tape and transaction tape unchanged. These original tapes could be rerun and a new master tape produced if a malfunction occurred during processing. This process clearly required a minimum of three tape drives with fast start/stop capability and the control facility to read from two tapes and write on the third tape simultaneously.

Magnetic tape was the primary file storage medium from the 1950s until the mid-1970s when disks, with their direct access, became large enough and reliable enough to provide a better answer. Although magnetic tape is rarely used for active data files in modern computer systems, it is still widely used for interchanging and archiving of data. But in the days before disks, tape played a central role in data processing. The big problem, of course, was that account balances were not available between posting runs. For example, services we take for granted today, like obtaining an up-to-the-minute bank balance, were not possible. Customers had to be content with data from a listing of the last batch posting run. This was a serious problem in many businesses. A problem that was destined to be the focus of a new laboratory and the development of the 350 RAMAC disk file.

Starting a New Laboratory

In late 1951 IBM decided to start a small new advanced development laboratory in San Jose, California. Reynold B. Johnson, veteran inventor from the Endicott Laboratory, was chosen to be the manager of the new San Jose operation. In a speech long afterward, Rey Johnson described his charter for the new laboratory:

> During the first week of January 1952 I was told of my appointment as West Coast Laboratory manager. I was told that I would have free rein in hiring a staff of 30 to 50, and I would be free to choose projects to work on. One half of my projects were to be new products and one-half were to be devices in support of customers' special engineering needs. No projects were to be duplicates of work in progress in other IBM laboratories. The laboratory was to be dedicated to innovation. To be given freedom to choose our projects and our staff made the San Jose Laboratory an exciting opportunity, especially since funding was guaranteed—at least for a few years.*

Most of the laboratory's early projects could be classified as input or output devices for punched-card accounting systems. On the input side, a main objective of punched-card systems was to get data into punch-cards as close to the source of its creation as possible. Several such "source recording" projects were initiated. A project for converting analog data from wind tunnel instrumentation (mostly strain gauges) into punched-card format occupied perhaps half the laboratory by the middle of 1952, and a tentative commitment, later canceled, was made to deliver some experimental hardware to Douglas Aircraft Corporation.

---

*Transcript of an address given at a dinner meeting of the Independent Disk Equipment Manufacturers Association (IDEMA).

The Source Recording Project and the Choice of Disks

The source recording project of interest here is one that envisioned replacement of a tub file of punched cards with some form of large-capacity magnetic storage. As we shall see, although many people in the San Jose lab initially thought it was folly to base that magnetic storage on a disk geometry, the decision gained enthusiastic supporters as solutions were found to such key problems as air-bearing support for the read/write head.

Keypunching was a costly item in a punched-card data processing operation, and this "source recording" project had the objective of increasing keypunch operator productivity by eliminating manual recording of fixed and repetitive information such as item descriptions, costs, names, and addresses. A keypunch operator would punch only variable data, say hours worked in a payroll application or quantity ordered in a wholesale application, with fixed data coming from some form of magnetic storage device attached to the keypunch. The questions were: What device? How big? How fast? and At what cost?

In September 1952 Art Critchlow and Geoff Hotham were assigned to study this problem. Magnetic drums and magnetic disks were the most likely candidates, but many other geometries were to be investigated. Critchlow and Hotham explored several rod and wire arrays, wide endless tapes on rollers, a system of drums, and a stack of magnetic plates. They also considered the notched disk array, described by Jacob Rabinow in the August 1952 issue of *Electrical Engineering:*

> Disks 20 inches in diameter are arranged on a common shaft that is bent into a circle so that the resulting envelope is a toroid or "doughnut." Each disk has a deep V-notch extending nearly to its center. The notches are aligned so that a magnetic head assembly can be rotated about the axis of the toroid and through the V-notch. A record is selected by rotating the head assembly to the desired disk and then rotating that disk so that it passes between the 64 recording heads which are on each side of the slot provided for the disks. Reading from the proper head provides for track selection. Access time is about one-half second. With recording densities of 13 tr/in and 100 b/in each disk could store about 500,000 bits.

On January 19, 1953, a significant simplification of the Rabinow scheme was tentatively chosen in preference to other proposed geometries. It envisioned a continuously rotating shaft containing an array of magnetic disks, with data access by moving a pair of magnetic heads to a selected disk and then radially on that selected disk to one selected recorded track among several concentric recorded tracks. The proposal for development of a disk storage device did not miss the final 350 by much, as can be seen in the following comparison. The 350 disk diameter was 24 inches, versus 16 inches for the initial proposal. Although the number of disks in the stack was 50 and the track density 20 t/in. as originally proposed, the bit density was half of that of the earlier proposal. The 1200 rpm speed and the 0.6-second seek time were close to the original 1953 proposal.

Continuously rotating magnetic disks were selected for a number of reasons in addition to the high surface-to-volume ratio they provide. Two functional advantages were achieved. First, the data were continuously available once the read/write head was positioned on a selected track, and one character per revolution could be read out to a keypunch with only a single character buffer. Second, multiple access could be provided by arranging several access mechanism around the array of disks.

The big problem with disks, however, was that axial runout (wobble) of as much as 0.1 inch was not uncommon, and no one knew how to maintain the close (< 0.001 in.) spacing required between the magnetic head and the disk surface in the presence of runout 100 times the required spacing. Although no answer to this major problem was known in April 1953, Rey Johnson reaffirmed his decision to go ahead with disks and discard all other alternatives. He was confident that the creative energies of his young laboratory would produce answers to this and other unknowns.

A Broader View of Disk Storage

Another reason for continuing the focus on disks was a growing realization that the applications of large-capacity random access disk storage were significantly broader and more important than an automated tub file. That vision had been stimulated by the receipt, in December 1952, of a request for proposal to provide an inventory control system for U.S. Air Force base supply. The system requested by the military would today be called an online database with remote access. It called for the storage and online access to 32,000 records of 20 digits each (640,000 digits total) that contained the inventory status of the most active airplane maintenance parts. Johnson assigned a small team of people to respond to the air force request, and a proposal submitted on March 31, 1953, specified 10 magnetic drums and a new small, drum-oriented, stored program computer. Disk storage was not yet ready for commitment to a customer delivery schedule.

Although the air force selected another contractor for the system, work on the proposal led to the notion that disk storage might become the center of a system vastly more important than anything considered to date. To understand this potential, John Haanstra was asked to take leadership in the preparation of a report on the possibility of associating a small computer with the disk file to provide a complete low-cost data processing system. This group was to survey applications, make a critical analysis of each, and postulate a machine. Haanstra's report of April 28, 1953, entitled "Preliminary Investigations of Applications for Random Access Memory to General Purpose Data Processing Systems," provided an early definition of online or random versus batch processing and gave a vision of the future online interactive applications made possible by disk storage.

The Disk Project Gets Moving on Some Hardware

Up to this point the work on disks had been study, analysis, and conjecture based on magnetic tape and magnetic drum technology; very little serious experimental work had been done. In April 1953, with staff available for new assignments, Johnson started making his vision come true.

The First Air Bearing

By the end of May 1953 there was good news. Bill Goddard had designed and "flown" a hydrostatic air bearing (Fig. 18–2) that maintained a constant close spacing between itself and the surface of a badly warped disk. And on June 2, 1953, a 51-bit record was successfully recorded and read back using an air-bearing-supported magnetic drum head and a spray-coated aluminum disk.

The attitude in the laboratory changed from one of skepticism to one of enthusiasm as this and the three other basic problems—disk substrate, disk magnetic coating, and access mechanism—were addressed and solutions found.

Disk Substrate

The first disks were turned from magnesium lithographer's metal, but this material was soon replaced by a special high finish aluminum alloy that Alcoa had developed for the substrate of master LP records. The stock material was only 0.051 inch thick; so after coating on one side, two disks were laminated, clamped between platens and, while under pressure, heated in a oven above the annealing temperature. The resulting disk was "dead" (no ringing) and fairly flat (runout of about 0.025 in.).

Disk Coating

The first successful magnetic paint was formulated by adding γ-Fe_2O_3 to a paint base similar to that used to paint the Golden Gate Bridge, but it was soon replaced by an epoxy base paint, which was used for many years for particulate disks. Initially the disks were spray-coated, but this technique could not give uniform surfaces. More uniform coatings were obtained by a spin-coating technique in which a measured amount of paint was applied at the center of a rotating disk and spread by centrifugal force. Spin coating was used for many years to coat particulate disks.

Figure 18–2 The first air bearing to "fly" at a small height above a wobbly disk surface. [*Courtesy of IBM Corporation.*]

Access Mechanism

After many alternatives had been considered, a clever access mechanism was devised by John Lynott. It used a common servo drive mechanism that was mechanically multiplexed to move a carriage from disk to disk at about 100 in./s, stop and detent at a selected disk, and then move the arm radially to a selected track location where the arm was detented to obtain the final position.

Work continued on the disk file (Fig. 18–3), and the file-to-card machine throughout 1953. On February 10, 1954, the first successful transfer of data from cards to disk and back again from disk to cards took place. However, by November 1953 it was already clear that the old file was not going to convince anyone outside the laboratory that disk storage could be an important new product. It was time to build a new model, and Rey Johnson asked me to take on the task.

Figure 18–3   The first disk file model (February 1954). [*Courtesy of IBM Corporation.*]

Designing the 350 RAMAC Disk Drive

In June 1954 work began on a second, improved, file design with Trigg Noyes handling the layout, John Lynnott the carriage and ways, Jack Harker the servo drive, Don Johnson the disk and disk coating, Wes Dickinson the servo electronics, Norm Vogel the air head and arm assembly, Al Hoagland and Ed Quade the magnetic read/write head, and Len Seader the read/write electronics and clocking circuits.

By November 1954, the first new file was operational and being tested around the clock. In this file the disk stack was made vertical to allow easier assembly and disassembly of the disks and to make room for several access mechanisms in a box small enough to go through a 3-foot door. Figure 18–4 illutrates how both disk-to-

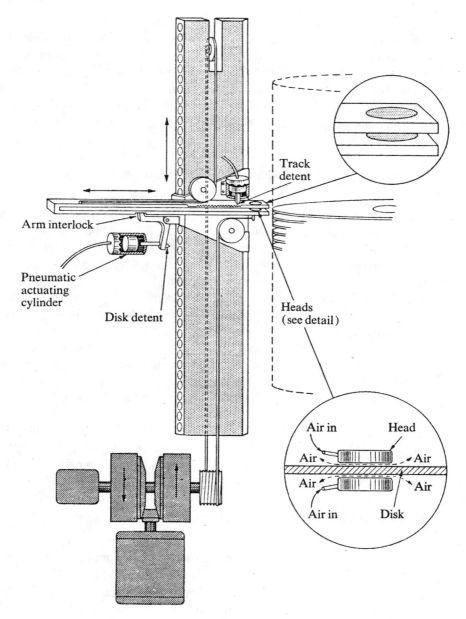

Figure 18–4 Schematic of the 350 disk file access mechanism. [*Courtesy of IBM Corporation.*]

disk movement and track-to-track movement were done with a single drive motor, with the final disk and track position determined by pneumatically actuated detents.

While all elements of the design were important, Norm Vogel's air head design was vital. Two small springs kept the head retracted into its socket while it was moving, and upon arrival at the destination track, air was turned on and three small pins acting as pistons (the head body was an annular manifold) pushed the head out of its socket toward the disk surface, while air escaping from the small holes on the front maintained the head about 800 μin. (20 μm) above the disk surface. The head contained two recording elements: one for writing and another with a narrower gap for reading. This write-wide, read-narrow scheme reduced the tolerance of final track positioning and was used in most subsequent products.

A Decision to Build Machines for Field Testing

With the new file operating well—over a million random read/write accesses had been completed—questions were being asked about the next step toward making it a product. The concerns of business for storage and processing data using magnetic tape were multiplied many times when a new and untried magnetic disk storage device was suggested. Practical operational procedures and safeguards to use such a device in business data processing were just not known, and Rey Johnson's suggestion of building several field test machines was seriously reviewed by corporate management. After considerable debate, the San Jose Laboratory was charged in November 1954 with designing and building 14 machines.

The disk file was in good shape but the computer part had not yet been started, although a preliminary specification had been prepared. Serious design started in December 1954 with first customer shipment scheduled for September 1955, leaving only nine months to design and build the first of 14 machines. This was a Herculean task for the relatively inexperienced staff of the small laboratory. Needless to say, the scheduled ship date of the first field test machine slipped. In June 1956, however, the 305-A RAMAC went out to the Zellerbach Paper Company in San Francisco.

RAMAC Is Headed for Announcement

By the fall of 1955 it had become clear from application studies and customer feedback, without any field test experience, that the RAMAC concept was practical. As a result, the 14-machine field test program became a real-world application development effort, and two big decisions were made. The first decision, in October 1955, was to start a redesign of the machine for production. After considerable debate as to whether to use vacuum tubes or the then new transistors, it was decided to stick with vacuum tubes; there were enough problems in getting the operation under way without the added burden of new technology. The second decision, in December 1955, was that production would take place in San Jose. This decision came after an extended debate on alternatives and was confirmed by purchase of a site in south San Jose for a manufacturing plant.

The hard disk industry, spawned by the 350 RAMAC, had annual revenues in 1997 that were conservatively estimated to be in excess of $40 billion.

THE EVOLUTION OF MAGNETIC DISK STORAGE

In this section we briefly discuss the characteristics of the trend-setting technologies and products that moved the hard disk drive industry in new directions. Some of these products were the first to use a seminal new technology; others used existing technology in a unique configuration, such as removable disk packs; and some used smaller diameter disks for reduced power and volume.

Self-Acting Hydrodynamic Sliders

A self-acting, hydrodynamic air bearing was first used in a disk file development begun in 1955 and known as the advanced disk file, or ADF. At first it envisioned the use of a slider-per-disk surface with a probe-type magnetic element that would record vertical flux transitions on a steel disk with a magnetite (Fe_3O_4) magnetic layer. The layer was formed by oxidizing the steel disk in a controlled steam atmosphere. After several years of effort, this technology was dropped for a number of reasons, including the difficulty of obtaining steel disks with acceptable defect levels.

Although the steel disk–vertical recording technology proved to be a failure, the ADF incorporated three innovations that are currently utilized in all moving-head disk drives:

- Self-acting hydrodynamic air-bearing (slider-bearing) support of the magnetic read/write head.
- An access arm composed of a comb of sliders with at least one per disk surface.
- "Cylinder mode operation" (i.e., for each radial position of the access arm, a head on any disk surface is selected by electronic switching).

These concepts and others were proposed by Jake Hagopian to be the basis of a "High Performance Magnetic Disk Random Access Memory." Hagopian's proposal was based on experimental work with slider bearings he had done in 1954 and later patented. The following quote, lightly edited from a June 28, 1954 entry in Hagopian's notebook is shown in holograph form in Figure 18–5.

Experiments were made today to demonstrate that it is possible to obtain air lubrication for a magnetic head riding against a disk without

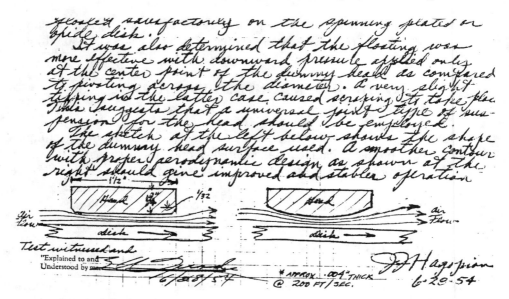

Figure 18–5   Excerpt from Jake Hagopian's notebook, including sketch of experimental slider. [*Courtesy of IBM Corporation.*]

the use of an air head and an air supply system. The horizontal turntable (12" disks) rotating at 1700 rpm was used, together with a dummy head made of aluminum cut 1.5 inch diameter and 0.25-inch thick. This dummy head was coated on one side with Everlube (molybdenum disulfide plus graphite). Both sides were lapped with fine emery cloth against flat stone before use.

With both plated and oxide-coated disks it was found that the Everlube provided a reduction in friction when the disk was rotating at very slow speed and there was actual physical contact between surfaces. However at higher speeds (above approx 100 rpm) actual air flotation of the dummy head took place because of the laminar film of air (approx. 004" thick @ 200 ft/sec) which forms and flows along the surface of the disk. It made no difference whether the Everlube-coated or the plain face of the dummy head was used. Either side floated satisfactorily on the spinning plated or oxide disk.

It was also determined that the floating was more effective with downward pressure applied only to the center point of the dummy head as compared to pivoting across the diameter. A very slight tipping in the latter case caused scraping to take place. This suggests that a universal joint suspension for the head should be employed. The sketch at the left below shows the shape of the dummy head surface used. A smoother contour with proper aerodynamic design as shown at the right should give improved and stabler performance.

Some of Hagopian's early successes were not consistently repeatable, and failure would occur when the slider momentarily touched the disk surface. With the 350-gram load required by these sliders, failure most often led to destruction of the disk surface, and the subsequent destruction of the magnetic head and slider. After some intensive analytical work by Bill Gross, and experimental work by Ken Haughton and Russ Brunner, it was found that reliable operation required an entry wedge for the boundary layer air film. The entry wedge was obtained by providing a cylindrical surface on the slider face much as Hagopian had intuitively predicted in his note book entry, but now it was known that the cylinder should have a radius of curvature of about 250 inches. In addition, Brunner and Haughton found that local disk flatness was important, and they established a specification of disk flatness in terms of velocity and accelerations of runout and local curvature.

Figure 18–6 shows an arm and head assembly as used in the 1301 disk file (1962). The 1301 drive consisted of two 28 MB storage modules, as in Figure 18–7, each with a stack of 25 disks, equipped with a comb of 50 head sliders positioned on the disk surfaces by a hydraulic actuator. The disks were rotated at 1800 rpm, and the average seeking time of the actuator was 165 ms. With a track density of 50 t/in. and a bit density of 520 b/in., the storage density achieved was about 13 times the 350 disk drive.

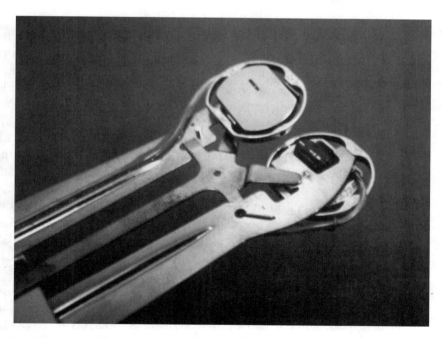

Figure 18–6 The slider suspension and load arm of the IBM 1301. [*Courtesy of IBM Corporation.*]

Figure 18–7   The 1301 disk file. [*Courtesy of IBM Corporation.*]

The Removable Disk Pack

We now review the evolution of a generation of disk drives with removable disks. These disk files combined the fast direct access of disks with the replaceable-medium advantage (and therefore very large capacity) of magnetic tape.

The 1311 was the first disk drive with a removable disk pack (Fig. 18–8). It began in early 1958 when a small team of engineers and planners were assigned the task of defining a "low-cost RAMAC system." While studying applications, Jack Harker, who was part of the team, recalls someone saying "If I could have disks that were removable like tape, I could use them in a mode like tape but skip over all of the inactive records during a file update and have the advantage of inquiry capability between file updates." Close inspection of this idea, which was called skip-sequential processing, verified that file maintenance would be substantially faster than with magnetic tape. Thus one of the recommendations of the study team's June 1958 report was the development of a low-cost disk drive with a removable and interchangeable disk. That recommendation was accepted, and Jack Harker was assigned to begin the work.

The first question Harker had to answer was whether removability was technically practical, considering the radial tolerances required of reading a removable disk years after it had been written on another drive. After considerable analysis he concluded that 50 t/in. was a reasonable objective. Furthermore, he decided that reducing the slider flying height to 125 μin. (3.1 μm) should permit

Figure 18–8 The IBM 1311 disk drive, with 1316 disk pack. [*Courtesy of IBM Corporation.*]

doubling the bit density to 1025 b/in. With these densities, a pack of six 13-inch (later increased to the long-time industry standard of 14-inch) diameter disks with 10 useful surfaces would have a capacity of 2 million characters, organized into 20,000 hundred-character records—the equivalent of 25,000 punched cards, or about one-fifth of a reel of tape. The 1311, shipped in 1963, achieved a track seeking time of 250 ms, using a simple two-speed hydraulic actuator.

A new business vista for disk storage was opened by the 1311 and broadened in 1965 by the 2311. The 2311, which increased the capacity of the 1316 disk pack to 7.25 MB, was designed for attachment to System/360 through a published standard interface, and by 1968 several vendors, including Memorex, Control Data, General Electric, and Potter Instrument, were offering plug-compatible versions of the 2311 disk drive, and more than a dozen vendors offered compatible 1316 disk packs.

The 2314 Storage Facility, first shipped in 1966, used a new disk pack, the 2316, which doubled the number of 14-inch disk surfaces from 10 to 20. The 2314 contained eight active disk drives and a spare, all managed by an common integrally packaged control unit. Each removable disk pack had a storage capacity of 29.2 MB, so the capacity of the facility with all eight packs mounted was 233 MB. The 2314 thus provided not only removable disk packs, for large offline capacity, but also was the first disk product with enough online capacity to encourage the development of the interactive online applications, taken for granted today.

Track-Following Servo

A significant new closed-loop head positioning technology in the form of a voice coil (Fig. 18–9) driven track-following servo was introduced in 1970. Open-loop voice coil actuators first appeared in 1965 on the 2310 "Ramkit," a removable single-disk cartridge file, and in 1968 on the 2311 plug-compatible Memorex 630. These early voice coil actuators used detents to determine track location. In a closed-loop voice coil servo actuator, the actual position of the heads is given by information recorded on a disk. This position information is compared with the desired position, and a position error signal is fed back to the voice coil to adjust the actual position. The actual position reference information was precisely recorded on a dedicated disk surface during manufacture. Most modern disk drives use a track-following servo but record the position reference information between data sectors rather than on a dedicated disk surface. In addition, voice coil actuators now use a rotary, rather than the linear mechanism of virtually all 14-inch disk drives, with the exception of the 1976 "Gulliver" disk drive, which used a rotary voice coil (open-loop) actuator.

In the 3330 disk drive, the servo system enabled more accurate positioning. This made possible a track density of 192 t/in. and an average seek time of 30 ms. In addition, the new rectangular ceramic slider (Fig. 18–10) reduced the flying height to 50 μin. (1.2 μm) and the glass-bonded ferrite magnetic head permitted a

Figure 18–9   The IBM 3330 voice coil actuator and head comb. [*Courtesy of IBM Corporation.*]

Figure 18–10   The IBM 3330 slider, suspension, and head. [*Courtesy of IBM Corporation.*]

bit density of 4040 b/in. giving an areal density of 0.776 Mb/in.^{2}—a 389-fold improvement over the 350 RAMAC.

Low-Mass, Lightly Loaded Sliders

The project that led to the development of the "Winchester" disk drive was started in the spring of 1969 as a repackaged, lower cost version of the 3330, aimed at providing a new file for small systems. Ken Haughton, then head of the advanced storage technology group, was asked to come up with a proposal for a product that would provide a disk drive with two spindles in one box, each spindle with a 30 MB disk pack. The box was known as the 30-30 and it was natural for Haughton to think of 30-30 in terms of a Winchester rifle; hence the project received the code name Winchester (Fig. 18–11). The resulting disk file, shipped in 1973, had a removable data module that contained either two or four disks (14-inch diameter), as well as the disk spindle and bearings, the carriage, and the head–arm assembly with two low-mass sliders per disk surface. This new design achieved an areal density of 1.69 Mb/in.^{2}, an increase of 845 times the 350 RAMAC.

Over time, "Winchester technology" became the industry-wide term for the combination of a lubricated disk and low-load ferrite sliders that are in contact with the disk when it is stopped but fly at about 20 μin. (500 nm) when the disk is

Figure 18–11   The IBM 3348 data module for the 3340 "Winchester" disk drive. [*Courtesy of IBM Corporation.*]

rotating. Haughton and his team synthesized the elegantly simple Winchester slider and head, shown in Figure 18–12, from two sources of experience.

The first source of experience was from a disk refresh buffer for a high-resolution display system that Joe Ma, in the Los Gatos laboratory, had designed,

Figure 18–12   The IBM "Winchester" suspension, slider, and head. [*Courtesy of IBM Corporation.*]

built, and installed at Rand Corporation. Ma had first tried 2314 heads on 24-inch disks but found them inadequate. After an extensive search, Ma located Data Disc Corporation, a disk drive start-up company in Palo Alto that was making an instant replay video buffer with a plated disk and a unique lightly loaded tripad head. In this design, a head had three small pads arranged in a triangle with sides of about 0.25 inch; the magnetic element at the apex served as one of the pads. The heads had a low mass and were lightly loaded with about 20 grams on a flexure suspension. Armin Miller, CEO of Data Disc and inventor of the tripad head (incidentally also inventor of the MFM or Miller modulation code), agreed to sell Ma a few tripad heads.

With little modification Ma was able to fly them at about 10 μin. (250 nm). His recording density problem was solved, but as a result of in-contact start and stop he was left with a wear problem on the 24-inch particulate disks he was using. Ma attacked the start/stop wear problem by spreading a thin layer of silicone oil on the disk surface. This solution worked well, and Joe Ma's disk became the first to use a lubricated particulate disk with a lightly loaded slider head. Haughton adopted the idea, but the tripad head was dropped because it was expensive to manufacture.

The second source of experience was from Erik Solyst, who had designed an easy-to-manufacture head for a fixed-head file, in which nine head elements and a flat slider with a taper at the leading edge were machined from a single block of ferrite. Taking the best from the two experiences Mike Warner designed the now ubiquitous Winchester head (Fig. 18–12). Low-load slider heads with lubricated disks became an important part of disk recording technology.

The low cost and simplicity of the head and slider structure made it practical to use two heads per disk surface. This design cut the stroke length of the voice coil actuator in half, decreasing access time and reducing the cost of the carriage and the actuator. In addition, the low-cost heads made a fixed-head option possible for customers who required faster access to some of their data at a reasonable cost.

Return to Fixed Disks

By 1975 most batch sequential processing had been replaced by online interactive database applications that required disk data to be available at all times, and so disk packs and data modules were no longer being used like tape reels. With the introduction of the 3350 fixed-disk drive design in 1976, things had come full circle, back to the fixed disk of early products. Now, with the return to fixed disks, the elimination of the tolerances that had been required to permit removability allowed higher recording densities. The 3350 used extended Winchester technology, with eight 14-inch disks per spindle, and achieved a capacity of 317.5 MB. Track density was increased to 478 t/in. and bit density to 6425 b/in., with areal density then of 3.07 Mb/in.^{2}.

Thin-Film Heads

The thin-film head, introduced 1979 in the 3370, represented a major advance over the magnetic heads of earlier disk drives. The development of the film head provided an improvement in areal density potential. Photolithographic and film deposition techniques similar to those developed for fabricating semiconductor integrated circuits, but with different materials, could be used to fabricate film heads with a new level of precision and control over the critical magnetic gap dimensions. In addition, the thin magnetic films provide the high-frequency performance needed as data rates increase with increasing bit density.

The first thin-film head (Fig. 18–13) consisted of several layers in a complex pattern of different materials deposited by photolithographic, etching, and electrodeposition techniques. The first layer is magnetic (Ni-Fe); it is shaped to be one of the poles of the head. Next comes a thin layer of insulating material, then the 8 turns of copper for the coil, more insulation, and finally the other magnetic pole layer. Many thin-film head elements are deposited simultaneously on a single ceramic substrate wafer about as thick as the length of the slider. These wafers are then diced and machined to yield individual sliders that contain the film head on the slider's trailing edge. This batch fabrication technique made the cost of high-precision film heads attractive. Figure 18–14 shows a pair of sliders with suspensions.

The 3370 had a capacity of 571 MB. It contained seven 14-inch-diameter disks that rotated at 2964 rpm and a media data rate of about 1.9 MB/s. With its thin film heads, 2,7 encoding, new and smaller sliders that reduced flying height

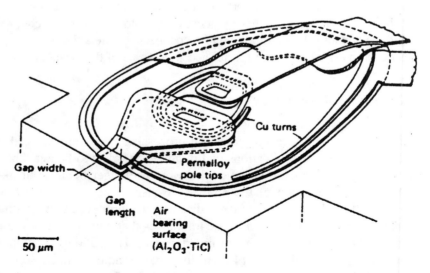

Figure 18–13   Perspective view of the IBM 3370 film head. [*Courtesy of IBM Corporation.*]

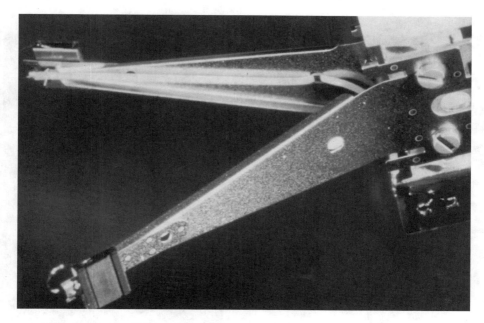

Figure 18–14   Slider and suspension for the IBM 3370. [*Courtesy of IBM Corporation.*]

to 13 μin. (320 nm), it achieved a bit density of 12,134 b/in. and a track density of 635 t/in., yielding an areal density of 7.7 Mb/in.^{2}.

The 3370 was the first of a generation of disk files with thin-film heads and led to a series of products achieving still higher capacity and performance. Three examples, the 3380, the 3380K, and the 3390-2, are described below.

The 3380 (1981) had a capacity of 1260 MB, contained nine 14-inch particulate disks, and rotated at 3620 rpm. With a slider and 8-turn film head similar to that of the 3370, it achieved a bit density of 15,240 b/in. and a track density of 801 t/in., for an areal density of 12.2 Mb/in.^{2}, a 6103-fold improvement over the 350 RAMAC.

The second product, the 3380K (1987), had a capacity of 3781 MB, in essentially the same mechanical assembly as the 3380. Its improved slider flew at an increased angle of attack, thus pressing the trailing edge closer to the disk and, using a new 31-turn film head, achieved a bit density of 17,200 b/in. With its all-digital servo system, a track density of 2089 t/in. was achieved, yielding an areal density of 35.9 Mb/in.^{2}.

The third product, the 3390-2 (1989), had a capacity of 3784 MB and contained nine 10.8-inch particulate disks. The smaller disks reduced power by more than a factor of 3, resulting in a sealed head-disk assembly of substantially smaller size and complexity than the 3380 (Fig. 18–15). The 3390, with a 31-turn film head, had a bit density of 27,940 b/in. and a track density of 2242 t/in., for an areal density of 62.6 Mb/in.^{2}, a 31,320-fold improvement over the 350 RAMAC.

Figure 18–15   IBM's 10-inch 3390 HDA (foreground) with equivalent parts from a 14-inch 3380 K. [*Courtesy of IBM Corporation.*]

Smaller Diameter Disks

The first move to smaller disks was in 1979, from 14 inch to 8 inch with the 3310. Next came the first 5.25-inch diameter disk drive, the Seagate ST 506, in 1980, and the first 3.5-inch drive, the Rodime RO 352, in 1983, and the first 2.5-inch drive, the Prairie Tek 220, in 1988. In this section we discuss this evolutionary trend to smaller diameter disks. The principal motivation for the trend is the limited space and power available in desktop personal computers that began to appear in the late 1970s, became ubiquitous, and are now shrinking to smaller, lighter, and truly portable sizes. We first review the interaction of disk diameter with other disk drive parameters and its impact on key components; finally we discuss the characteristics of the first 8-inch and 5.25-inch hard disk drives.

Many of the important parameters of a magnetic disk drive are interdependent. That is, a change in one will result in a change in another parameter or parameters. For example, a change in disk diameter is reflected as a change in power, capacity, and data rate.

The power required to rotate a disk increases rapidly with rotational velocity (rpm) and even more rapidly with the disk diameter (d) as given by the equation:

$$power = (constant)(rpm)^{2.8}(d)^{4.6}$$

For a given rpm, a 14-inch disk will take 13 times the power of an 8-inch disk, 91 times the power of a 5.25-inch disk, 588 times the power of a 3.5-inch disk, and 2765 times the power of a 2.5-inch disk.

The capacity, in megabytes (MB), of a disk is proportional to the square of the disk diameter and the areal density capability of the magnetic recording technology.

$$MB = (constant)(b/in.)(t/in.)(d^2)$$

So for a given areal density a 14-inch disk has 3 times the capacity of an 8-inch disk, 7 times the capacity of a 5.25-inch disk, 16 times the capacity of a 3.5-inch disk, and 31 times the capacity of a 2.5-inch disk.

The media data rate is a linear function of disk diameter, bit density, and rpm. That is, if you double the rpm, the bit density, or the diameter you double the data rate.

To reduce rotational latency and data transfer time, we would like to rotate the disks at the highest rpm possible. At high bit density, however, the disk rpm will be limited by either the ability of the host computer to handle the resulting high data rate or the power dissipated by the disk array as an inefficient air pump.

Smaller diameter disks have made a significant impact on the disk itself by making sputtered alloy film disks practical and thus permitting retirement of the γ-Fe_2O_3 particulate disk coating used for the first 25 years in essentially all hard disk products. Particulate disks were limited in areal density potential by their relatively poor magnetic properties and physical surfaces. They were replaced by the end of the 1980s by sputtered thin magnetic alloy film disks whose superior magnetic properties and smoother surfaces permitted flying height to be reduced to several microinches today with attendant increases in density. Further discussion of sputtered thin-film disks follows in the next section.

The first 8-inch hard disk drive, the 3310, was originally designed as a subassembly for integration into the package of small, low-cost computers. Small size, low power consumption, and high reliability were prime design requirements. The heat generated by this small 8-inch disk drive was low enough to be dissipated directly through the sealed disk enclosure, thus providing low-cost protection without requiring the extensive external air cooling systems required by 14-inch disk drives. The capacity of the drive was 64.5 MB. Its disk enclosure contained six 8-inch disks with one surface dedicated to servo position data. It used Winchester heads similar to the 3350 and achieved a bit density of 8,530 b/in., a track density of 450 t/in., and an areal density of 3.8 Mb/in.^{2}.

Although rotary actuators were considered as far back as Hagopian's 1954 "high performance magnetic disk random access memory," and used on the single 14-inch-disk Gulliver drive in 1976, they were not common with large-diameter disks. The 3310 appears to be the first 8-inch production disk drive with a rotary voice coil actuator. Virtually all modern disk drives use rotary voice coil actuators. In addition, the 3310 appears to be the first production disk drive to use a sector servo to provide track following. The system was a hybrid scheme that provided an average seek time of 27 ms. During seek operations the dedicated servo surface provides position information. When the desired cylinder is reached, however, its position information is derived from the selected data head reading prerecorded feedback information from between data sectors. Sector servo position control eliminates temperature effects and most of the tolerances between the dedicated servo head and the data head. Essentially, all modern small disk drives use sector servos for seeking as well as track following.

The first 5.25-inch hard disk drive was the Seagate Technology ST 506 (1980). Seagate was founded in 1979 by Alan Shugart with the specific objective of using Winchester technology to provide a small 5 MB hard disk as a physical and electrical upgrade of the 5.25-inch floppy.

The ST 506, with its capacity of 5 MB—huge in comparison with a 360 kB floppy disk—was designed to sell for under $1000. It was a great boon to early personal computer users, who could easily replace one of the two 5.25-inch floppy disk drives in a system and immediately reap the rewards of 15 times higher capacity online storage and substantial improvement in performance. The ST 506 was a great success, and many other manufacturers entered the market. By the end of 1982 more than 200,000 5.25-inch Winchester drives had been shipped by a new 5.25-inch Winchester industry that numbered 40 manufacturers, accounting for 125 different 5.25-inch Winchester products. Seagate is now the world's largest independent manufacturer of hard disk drives, and 5.25-inch drive capacities have risen to 18 GB.

The ST506 used a stepper motor similar to that in a floppy disk drive to linearly move an arm with four Winchester heads. Its step time was 3 ms per cylinder, and average seek time was 170 ms. It achieved 5 MB on two 5.25-inch particulate disks, recording at a bit density of 7690 b/in. and a track density of 255 t/in., for an areal density of 1.96 Mb/in.^{2}, a 980-fold improvement over the 350 RAMAC.

Magnetoresistive Read Heads, Thin-Film Disks, and PRML Channels

This section briefly discusses the impact of three technologies introduced in disk drive products in the late 1980s and early 1990s: the magnetoresistive read head, (MR head) the high-coercivity, thin-film disk, and the partial response, maximum likelihood (PRML) channel. In addition to reduced flying height, they contributed significantly to the more than 65% growth rate in areal density in the 1990s shown in Figure 18–1.

From their introduction in 1979, thin-film heads and improved, smaller slider technology permitted significant increases in areal density by reduction and better control of the critical head and spacing dimensions. However, by the end of the 1980s it was clear that the difficulty of achieving adequate signal-to-noise ratio would limit future improvements, especially in track density, since the signal from an inductive read head is proportional to the rate of change of flux, and thus a function of the track width, number of head turns, and disk linear velocity (rpm and disk diameter). The tactic of increasing signal strength by increasing the number of turns (from 8 turns in 1979 to 37 turns in 1991) as track widths were reduced and track density increased (from 635 t/in. to 2242 t/in.) had nearly reached its limit. The signal-to-noise problem with inductive film read heads was made more difficult with smaller diameter disks with their lower linear velocity and thus lower signal. The alternative is to replace the inductive read head with a flux-sensing head. In such a head, the recorded magnetization is sensed directly, and the readback signal is independent of the speed of the disk. The most successful flux-sensing-head approach has been the magnetoresistive (MR) head.

The MR head is based on the phenomenon of magnetoresistance (i.e., the variation in resistance of certain metals in the presence of a magnetic field). The phenomenon is old, having been discovered in 1856 by Lord Kelvin. Robert Hunt at Ampex invented the magnetoresistive read head in 1971, and IBM first used the technology in 1985 in magnetic tape head. The first disk product to use an MR head was the IBM Corsair (1991). This drive contained eight 3.5-inch film disks with a capacity of 1 GB. The bit density was 58,874 b/in. and the track density was 2238 t/in. for an areal density of 89.54 Mb/in.[2].

A magnetoresistive read element can be combined with an inductive write head. Such a dual-element head design permits the use of the inductive head for writing and the magnetoresistive head for reading.

Figure 18–16 shows a thin-film head on a microslider (micro when compared with a minislider, as used in a Winchester drive) and an MR head with its even smaller nanoslider. An MR head is shown in Figure 18–17.

Thin-film disks are a multiple-layer sandwich of thin films. They start with a thin substrate of Al-Mg that has been diamond-turned or fine-ground then plated with a 10 μm thick Ni-P sublayer. This hard nickel layer allows polishing down to 5 μm, resulting in a nearly perfect disk substrate surface, free from pits and asperities. Onto this polished sublayer are sputtered a 50 nm underlayer of Cr and a 30 nm magnetic layer of cobalt alloy (Co-Cr-Ta, Co-Pt-Cr, or Co-Pt-Ni). The magnetic alloy and the sputtering conditions allow a coercivity greater than 3000 Oe today. Then a sputtered overcoat of 10 to 15 nm of "diamond-like" carbon provides a protective layer and, finally, a very thin film (2–3 μm) of lubricant is applied to the disk surface. By the end of the 1980s this technology, pioneered in

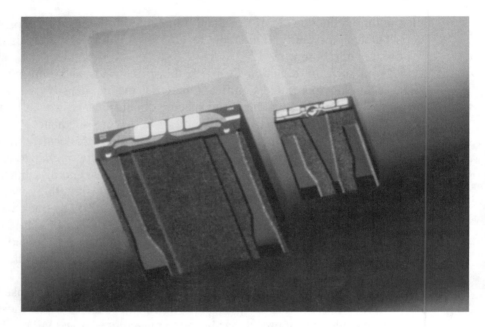

Figure 18–16   Left, film head and microslider; right, MR head and nanoslider. [*Courtesy of IBM Corporation.*]

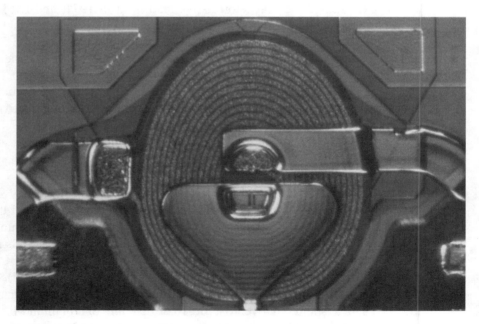

Figure 18–17   Film inductive write head combined with MR read head. [*Courtesy of IBM Corporation.*]

Figure 18–18 The IBM Travelstar 5 GT has the equivalent capacity of all the 350 disk files produced, and it is small enough to fit in a shirt pocket. [*Courtesy of IBM Corporation.*]

the 1980s by Lanx, Seagate, IMI, Maxtor, CDC, IBM, and others, was the preferred disk for small diameter drives; as large-diameter drives phased out, it became the universal design. Thin-film disks, with their cleaner head–medium interface, allowed reduction of flying height and thus contributed significantly to increasing potential areal density.

The PRML channel is not dependent on the MR head or the thin-film disk. It was first used with a ferrite head and has replaced the peak sense channel, in use since 1966, and has had an important but second-order effect in improving areal density. The PRML channel is a variation on data communication technology that samples an analog signal (from the read head), converts it to digital, and uses digital techniques to transmit and process the signal. At the receiver the data is processed according to the Viterbi (after its inventor) algorithm, which selects the most likely group of bits from all possible groups of bits. The analogy between the magnetic recording channel and a data communication channel is obvious, and the advantages of sampled detection in a noisy environment are well known in data communications.

SUMMARY

In this chapter we have traced the history of hard disk storage products and their technology from conception through gestation, birth, and growth. We have seen that growth is defined in this technology not as becoming larger but as shrinking, to get more storage capacity in a smaller package at lower cost per megabyte. One model of the 1997 2.5-inch Travelstar series of disk drives is shown in Figure 18–18. It is designed for laptop computers and other mobile applications and contains three 2.5-inch disks with a capacity of 4 GB. It achieves 211,000 b/in. and 12,500 b/in. for an areal density of 2.64 Gb/in.^{2}, a 1,381, 750-fold improvement over the 350 RAMAC. The other model contains four disks with a capacity of 5.1 GB, which is about the same capacity as all the 350 RAMACs shipped between 1957 and 1961.

Hard magnetic disk drives could be purchased at retail in 1997 for less than 5 cents per megabyte, as contrasted with the 1957 cost of about $10,000 per megabyte for the 350 RAMAC. Enough said!

REFERENCES

Ashar, Kanu, *Magnetic Disk Drive Technology,* IEEE Press, Piscataway, NJ, 1997.

Bashe, C. J., L. R. Johnson, J. H. Palmer, and Emerson W. Pugh, *IBM's Early Computers,* MIT Press, Cambridge, MA, 1986.

Carey, John, "Magnetic Field of Dreams," *Business Week*, 118–123, April 18, 1994.

Chiu, A., et al., "Thin-Film Inductive Heads," *IBM Journal of Research & Development,* **40**, 283–300 (1996).

Comstock, R.L., and M. L. Workman, "Data Storage on Rigid Disks," in *Magnetic Storage Handbook*, C. D. Mee and E. D. Daniel, Eds., 2nd ed., McGraw-Hill, New York, and IEEE Press, Piscataway, NJ, 1996, pp. 2.1–2.134.

Conner, Finnis, "Introducing the Micro-Winchester," *Mini-Micro Systems*, 79–82, April 1980.

Hagopian, J. J., "Data Storage Apparatus," U.S. Patent 3,007,144 (1961).

Harker, J. M., et al., "A Quarter Century of Disk File Innovation, "*IBM Journal of Research & Development,* **25**, 677–689 (1981).

Hunt, R., "A Magnetoresistive Readout Transducer," *IEEE Transactions on Magnetics,* **MAG-7**, 150 (1971).

IBM Corporation, "Disk Storage Technology," GA26-1665, IBM, San Jose, CA, 1980.

Kean, D. W., *IBM San Jose: A Quarter Century of Innovation*, IBM, San Jose, CA, 1977.

Noyes, T., and W. E. Dickinson, "The Random Access Memory Accounting Machine—II. The Magnetic Disk Random Access Memory," *IBM Journal of Research & Development,* **1**, 72–75 (1957).

Pugh, Emerson W., Lyle R. Johnson, and John H. Palmer, *IBM's 360 and Early 370 Systems,* MIT Press, Cambridge, MA, 1991.

Rabinow, J., "The Notched-Disk Memory," *Electrical Engineering,* **71**, 745–749 (1952).

Warner, M. W., "Flying Magnetic Transducer Assembly Having Three Rails," U.S. Patent 3,823,416 (1974).

Wolf, Michael F., "The R & D 'Bootleggers': Inventing Against the Odds," *IEEE Spectrum*, 38, July 1975.

19 Data Storage on Floppy Disks

David L. Noble

In the early 1940s, the punched card was the primary source of information input and output, and eight different types of punched card machines were in use: collator, interpreter, keypunch, multiplying punch, reproducer, sorter, tabulator, and verifier. These machines revolutionized many paperwork tasks. Also, printers had been developed for printing reports and documents as well as printing on the card itself for the interpretation of its punched columns. Cards were keypunched in much the same way everywhere.

In 1973 a new information communication system appeared consisting of a keyboard station, a cathode ray tube monitor, and a magnetic recording medium called a diskette. The diskette, a small flexible disk in a protective jacket, significantly altered the method of data entry and led to the demise of the punched card.

In 1977 the personal computer was first mass-produced. The 5.25-inch diskette played a very important role in the acceptance of the personal computer in the home and the office by offering a large amount of removable nonvolatile storage. In seconds, a diskette can be ejected from its drive and replaced by another containing alternate programs, text, or data. Search and access to specific information from the diskette is direct and easily updated. As with magnetic tape, it is possible to store recorded material for years. The diskette is portable, inexpensive, and reusable, and it can be read anywhere.

This is the story of the history of the development of the "little disk that could."

COMPUTER OPERATIONS

In the early 1960s it became possible to enter information, submit requests, and receive results via "terminals" located at a distance from the computer itself. A terminal is a device composed of a keyboard for putting data into a computer, and a video screen or printer for receiving data from the computer. A basic electronic computer system was comprised of at least five major sections: input, memory, an arithmetic–logic unit (ALU), output, and control (see Fig. 19–1). The input section consisted of input equipment for feeding instructions and/or original data into memory. Input devices included terminals, card readers, magnetic tape drives, optical scanners and, depending on the application, other equipment such as magnetically inscribed check readers, paper tape readers, and scanners for computer process control. The memory section stored programs and the initial, intermediate, and final processed data. Output included equipment for receiving results from memory in a form that could be either read by users, recorded, or used by other computers for further processing. These devices were printers, card punches, magnetic tape drives, cathode ray monitors, and equipment for transmitting data over telephone lines. The ALU performed the required addition, multiplication, division, and logic operations. The control section, the "brains" of the

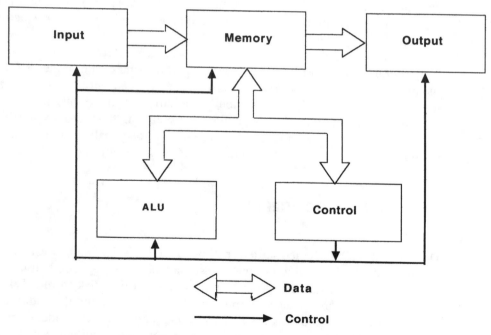

Figure 19–1   The basic organization of a computer.

system, was a collection of memory registers and circuits that accessed the program, deciphered data, and orchestrated the operation of all the other sections of the computer system. It did this by means of a read-only set of fixed instructions called microcode. In the 1960s, high-speed main memory consisted of doughnut-shaped ferrite cores, strung with wires and mounted in a three-dimensional array so that a given core could be selected and magnetized in one direction or in the opposite direction. Cores retained their states even after the computer power had been interrupted or turned off.

In 1967 IBM made a bold strategic decision for its new product line, the System 370. This decision was to discontinue the manufacture and further development of magnetic cores in favor of monolithic semiconductor memory chips. The reason was the promise of shorter memory access time and a reduction in manufacturing costs. The plan also included a change in control. The microcode, which had been stored in an extension of the main memory, was to become easily changeable. However, when the power to a monolithic semiconductor memory is interrupted, all stored information is lost. System designers therefore planned to introduce a separate, nonvolatile device for loading microcode directly into control. When the power was turned on, microcode could be automatically loaded into the control store from the new device, providing a method of introducing initial microcode instructions for getting the computer system up and running. Also the use of alternate sets of microcode would be advantageous when a computer system was upgraded, for routine computer checkout before a run, and for supplying the customer engineer with alternate diagnostic microcode for isolating system problems.

Among the specifications for this new device, referred to as an initial control program load (ICPL) device were a manufacturing cost of under $200 for the medium reader and a factory-written replaceable read-only medium that could hold 256 kb to 1.024 Mb of microcode. In addition, the read-only medium had to be easy to ship, and it had to cost under $5. Foremost, the reliability of the device and medium had to be unquestionable, since a load malfunction or data error would result in a completely inoperative computer system.

FUNDAMENTAL CONCEPTS

Design of the ICPL device started in November 1967 with an analysis of various recording media in common usage. These included the magnetic belt used in the IBM dictating machine, the grooved magnetic recording disk from a Telefunken dictating device, the standard 45 rpm phonograph record, and the audio tape cartridge. Also considered was a small rigid disk and a stretched-membrane magnetic disk under development in the laboratory. None was fully acceptable for the task; however, combining some of the features of these devices was a beginning.

A flexible disk was selected as the medium for in-contact magnetic recording. The original concept was to stamp disks from 1.5-mil magnetic tape material. Particle orientation of the tape coating was to be left random rather than longitudinal, as in the manufacturing practice for computer tapes. The center of the disk was to have a 1.5-inch-diameter hole so that, after insertion into a disk drive, a self-centering flange could clamp the disk to a vertical rotating round table. For stiffness and for recording head-to-disk compliance, a 1/8-inch foam pad was affixed to the back of the disk. The disk would also contain eight 0.1-inch-diameter holes, equally spaced just inside its outer boundary. These holes, sensed by a photodetector, would serve as sector markers to reference the location on the disk of the beginning of an individual record.

Frequency modulation (FM) recording was selected for in-contact recording. Using this method, the start of each recording cell was initiated by a flux reversal. This was referred to as the clock bit and provided a self-clocking system. Data bits were written between the clock bits. A 1 would be encoded as a flux reversal. A 0 would be encoded as a non–flux reversal. When recovering recording data, the clock bit opened a window and a recording detector looked for the presence or absence of a data bit. Laboratory testing using this recording arrangement indicated that with an inner track diameter of 4 inches, a linear density of 1600 b/in. was attainable. From this it was determined that a 7.5-inch diameter disk could support a 1-inch recording band and the eight sector markers.

Product testing for contaminants, together with an analysis of proposed disk mailing envelopes, led to the idea of rotating the magnetic disk inside a permanently enclosed envelope or jacket. The jacket eliminated the need for the disk foam pad, and the drive turntable could be replaced by a rotating hub. The jacket was fabricated from two 8-inch-square vinyl cover sheets separated by a vinyl spacer sheet containing a large aperture for acceptance of the disk. An adhesive was used to assemble the sheets. Head-to-disk contact was accomplished through elliptical apertures, one on each side of the jacket. A spring-loaded pressure pad was placed on the side of the disk opposite to the head to press the disk surface against the head. Also, matching holes through the jacket were provided for the disk-clamping system and for the sector-sensing system (see Fig. 19–2). The inside covers of the jacket were lined with a non-woven, antistatic fabric, which served as a continually wiping surface to cleanse the disk of contaminants as it turned. Most important, the antistatic characteristic of the wiping material served to prevent static electricity from being generated between the rubbing surfaces. Static charge would not only hinder disk rotation, but would accumulate and then discharge via the head and through the sensitive magnetic recording amplifiers, thereby introducing noise and causing read errors.*

---

*A patent for a magnetic record disk cover for combination of recording disk and jacket for containing the disk was granted to IBM Corporation on June 6, 1972 (U.S. Patent 3, 668, 658).

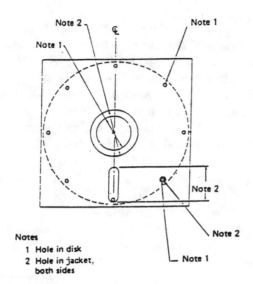

Notes
1 Hole in disk
2 Hole in jacket,
   both sides

Figure 19–2 Schematic of an 8-inch-square vinyl protective magnetic disk jacket indicating apertures for disk drive, head access, and optical sector markers. [*Courtesy of IEEE Transactions on Magnectics.*]

The disk, within its new enclosure and without its foam pad, was strengthened by doubling its substrate thickness to 3 mils. Although information was to be recorded on one side only, both sides were coated with magnetic material to provide identical sliding surfaces. A write-wide, read-narrow, recording/reproducing arrangement was selected to overcome the various mechanical and environmental tolerances the device was expected to encounter. Writing a 32-mil-wide track under controlled conditions at the factory, and reading with a 12-mil head in the field, would suffice for 32 tracks over the 1-inch recording band. Metal core tape heads were modified for factory writers and disk readers. Head carriage access to individual tracks was accomplished by means of a lead screw driven by a rotary solenoid. Track-to-track access time was 333 ms.

The disk speed had been established at 90 rpm. Experiments were conducted to be sure that the recording heads had an optimum gap-to-disk separation across the variable velocity recording tracks, and the resulting data led to the selection of a 2.5-inch spherical radius for the face of the heads. Experimental work also indicated that disk signal quality could be improved by a burnishing process. Burnishing involved passing a cutting edge across the recording band to reduce surface roughness and remove high spots. This process not only improved the readback signal envelope but also decreased disk defects, thus improving disk yield.

Based on 32 tracks at 1600 b/in., the formatted capacity of the disk was 81.6 kB. Formatted capacity is the total number of bytes that can fit on a disk surface not including space lost defining boundaries between sectors. The ICPL device was incorporated in the control unit of the early System/370 processors and also in the control unit for the IBM 3330 rigid disk storage drives, with customer deliveries beginning in 1971. The flexible disk drive was given the internal product designation of 23-FD, and the disks were designated as "diskettes." The press, however, called the diskettes "floppy disks."

THE WRITE REQUIREMENT

During the final stages of the development of the read-only initial control program load device, the Field Engineering Development Group needed a read/write version. Plans were being formulated to install an autonomous diagnostic processor in future System/370 processors, and the ability to capture system errors (usually corrected by error codes or instruction-retry), monitor selected logic points, and record status conditions would be an advantage for system serviceability. The flexible disk file, with the capability of online recording, would be the ideal device for maintenance logging. Proposals were also under way for a follow-on flexible read/write disk file with increased capacity and improved access time.

Laboratory recording trials indicated that a reduction in the magnetic coating thickness from 150 to 100 µin. (3.75 µm to 2.5 µm) would permit the achievement of a maximum recording density on the innermost track of 3200 b/in. Surface burnishing of the disk was again used, but with finer tolerances to prevent damage to the coating by removal of too much material.

As a reliable information exchange, any disk written under worst-case environmental and mechanical tolerances must be readable within a large population of other drives, each operating within a maximum tolerance range. Off-track recovery of recorded signals was the first hurdle to be addressed. The answer was to employ a "tunnel-erase" pole-tip geometry. Here, a new track is written wide and trimmed narrow, with tunnel-erase gaps that obliterate either side of the new track and neutralize previously written and unwanted information. Next, head carriage track-to-track accessing accuracy was substantially improved by incorporating a Geneva mechanism between the lead screw and a 90-degree stepper motor. Data from detailed studies of heads off-track together with the other tolerances were collected and tabulated. Computer Monte Carlo simulations using these data predicted the probability of a head operating beyond its tracking limits. From the results of these studies it was determined that the track density could be increased from 32 t/in. to 48 t/in.

The recording band of the disk was enlarged by increasing the disk diameter from 7.5 inches to 7.88 inches, and replacing the eight sector holes by a single index hole located inside the innermost track. Sector addresses were replaced by this index indicating the beginning of a track, and individual sectors were labeled by magnetic recording markers on the track. The larger recording area permitted the number of tracks of the original 23-FD to be increased from 32 to 77. The disk formatted storage capacity was now 243 kB.

Track-to-track access time of the head was improved from 333 ms to 50 ms by the stepper motor and Geneva mechanism. Access time was also improved by increasing the disk rotational speed from 90 rpm to 360 rpm. The higher surface speed required additional studies be performed on the head-to-disk compliance, and resulted in a reduction of the spherical head radius from 2.5 to 2 inches. A compatible lubricant was found to offset head and disk wear caused by the higher disk surface velocity.

With the increase in both rotational speed and linear bit density, the fixed timing window of the original data recovery system was no longer reliable. This was a result of phase shifts of the sensed flux reversals. To adjust for phase shifts, a variable frequency oscillator was added. The oscillator was synchronized to the frequency of the data being read, and special data were encoded on the track for this purpose. This arrangement dynamically adjusted the timing window to compensate for position variations of the recovered signal.

To reduce manufacturing cost, the jacket was now stamped from sheets of a copolymer material. The stamping also included the matching apertures for the self-centering disk clamp, the read/write head, and the optical sector detection system. The wiping fabric was heat-sealed to the inner surface, the one-piece assembly folded, the recording disk placed between the fold, and the jacket tabs on the other three sides folded over and heat set. The completed jacket dimensions remained 8 inches on a side. As a further cost reduction, the metal and tunnel-erase cores of the read/write head were replaced with cores of ferrite material.

The IBM 3740 Data Entry Station, delivered to customers in 1973, was the first product to use a read/write diskette and drive, the 33-FD. The station was equipped with a keyboard and a visual display. It had the capability of direct access to diskette records. It could also read from or write to standard magnetic tape, or to or from an IBM System/370 processor.

An inexpensive, convenient storage and exchange medium, the diskette had the capacity of about 3000 80-column punched cards. Eventually it replaced cards, card readers, and card punches, which had been in vogue since the 1890 census. In 1977, nineteen companies were manufacturing floppy disk drives in the United States. Their combined worldwide output had exceeded 200,000 units per year and was growing rapidly.

Following the acceptance of the diskette as a new medium for data exchange, new applications revealed the need for additional flexible disk capacity. Other applications would be extensions of original systems. The next iteration of the flexible disk drive was therefore to be downward compatible, that is, to be able to read and write disks processed on the 33-FD in addition to follow-on diskettes.

In 1976 a drive with two heads (43-FD) was available for recording on both sides of the diskette, thereby doubling its formatted storage capacity to 568 kB. Placing two opposing heads on either side of the disk created compliance problems. The design problem of the two heads and their carriage was not trivial. Influenced by the flying-head technology developed for rigid disk storage units, the flexible disk designers selected a flat surface head with shallow channels on either side of the read/write–erase cores. Opposing heads were symmetrical to each other; however, to prevent magnetic coupling through the disk, the opposing cores were offset by four tracks. Care in the design of channel dimensions and the outer edges of the flat surfaces was essential to maintain a suitable air bearing beneath the head to ensure signal stability. The head mounting system consisted of a pair of symmetrical head suspension arms with prestressed load springs, which pressed the head against the disk until the subsequent application of a head unload force. A diskette removal operation caused the arms to further separate the heads. Also, the head carriage lead screw, Geneva mechanism, and stepper motor were replaced by a set of guide rails and a band–capstan–stepper motor combination. This reduced the track-to-track access time from 50 ms to 5 ms. To distinguish between diskettes processed by 33-FD and 43-FD drives, the index aperture in the 43-FD jacket was located at a different position. Two separate optical systems were used. The two types of diskette were designated Type 1 and Type 2 for the 33-FD and the 43-FD, respectively.

In 1977 the bit density of the diskette was again doubled (53-FD). This was accomplished by a data encoding system called modified frequency modulation (MFM). For a stream of data to be recorded, magnetization is reversed at the beginning of a bit cell boundary only when data zeros are present in the preceding and present bit cells. Otherwise a reversal in magnetization is made at the center of a bit cell time for each data cell containing a 1. The maximum magnetization reversal rate remains the same. However, by utilizing successive data cell information pairs, the encoding permits the amount of recording information to be doubled. To distinguish between diskettes processed by 33-FD and 43-FD systems, the index aperture in the jacket of the 53-FD was moved to a third position. For downward compatibility of the three types of diskette, the location of the index hole is used to switch the read/write electronics for alternate recording modes.

Expanding small businesses making the transition from paper to automated accounting employed the IBM 5110 and 5120 minicomputers utilizing the direct access storage performance of the 53-FD diskette drive.

Systems with nonremovable direct access storage drives often used diskettes for archive and backup storage. As the size of the system storage increased, it was necessary to use a series of diskettes for this purpose. Now users required the ability to process groups of diskettes automatically with limited operator intervention. The specification for the next flexible disk drive called for the capability of processing an assemblage of the three diskettes types arranged in any order. In 1979 IBM made available a diskette magazine drive (the 72-MD). This drive was engineered to process automatically multiple diskettes under system control. A diskette magazine was designed for retaining up to 10 disks. Two magazines could be placed into the drive at any one time. Also provided were positions for insertion of up to three individual diskettes. In operation, a load mechanism would grasp a specified diskette and place it in the drive for read/write processing, then replace it back in its magazine and move another diskette into the read/write position. Adjacent diskettes could be exchanged within 3 seconds. With two magazines in place, the maximum attainable formatted capacity was 24 MB. The drive was also engineered to rotate the diskette at 720 rpm, twice the speed of the other drives. This required a reengineering of the read/write electronics to compensate for the change in disk velocity. The diskette magazine drive provided a significant enhancement of the diskette for large amounts of data storage, retrieval, and/or revision. The IBM systems 32, 34, and 38 utilized the magazine drive for initial control program load, diagnostics, and archive storage, and as an alternate for loading the computer operating system program.

Since the introduction of the diskette as a medium of information exchange in 1971, many other companies became engaged in the development of flexible disk drives. In 1976 Shugart Associates made available a 5.25-inch flexible disk drive that had recording characteristics similar to those of the 8-inch design in use at that time. Because of the smaller disk diameter, the bit density was limited to 2581 b/in. The track density was 48 t/in., and the total number of recording tracks was 35. The formatted storage capacity of this diskette was 81 kB. Track-to-track access was accomplished by a stepping motor driving a positive motion spiral cam mechanism. The access time was 40 ms, and the rotational speed was 300 rpm. This 5.25-inch floppy drive was a significant enabler for early personal computers.

As was the case in the 8-inch drive, a second head was later added to use both sides of the disk. The MFM encoding scheme was used to boost the bit density to 5922 b/in. Because the hygroscopic and thermal expansion of a 5.25-inch disk is less than that of an 8-inch disk, it was possible to increase the track density from 48 t/in. to 96 t/in. and to enlarge the total number of tracks to 160. The formatted disk capacity was now 655 kB. Track-to-track access time was improved to 3 ms. The rotational speed remained at 300 rpm.

In 1981 the linear bit density of the 5.25-inch disk was increased to 9646 b/in. by changing the coating material to cobalt-modified ferric oxide and reducing the coating thickness to 1.2 μm. The final formatted storage capacity became 1262 kB. The track-to-track access time remained the same; however, the rotational speed of the drive was changed from 300 rpm to 360 rpm.

In 1981 Sony started production of a 3.5-inch drive and diskette. The disk was placed in a hard plastic jacket that did not bend. The jacket incorporated a nonmagnetic metal shutter arrangement at one edge to provide an opening to the disk surface for entrance of the recording head, while further protecting the disk from contamination. The small diameter and the use of a metal hub bonded to the disk center reduced disk positioning runout tolerance, allowing a track density of 135 t/in. and 80 tracks on one surface. The linear bit density was 8717 b/in. and the formatted capacity was 322 kB. Rotation of the disk was 300 rpm. Track-to-track access time was 3 ms. In time, a second head was added to the opposite side

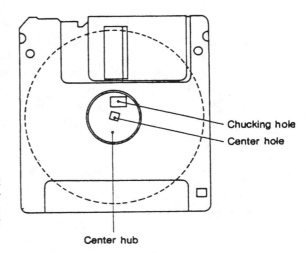

Chucking hole
Center hole

Center hub

Figure 19–3 Hard plastic 3.5-inch disk cartridge showing shutter that closes over the head-accessing window to protect the disk surface. [*Courtesy of McGraw-Hill and IEEE Press.*]

of the disk and the total number of tracks increased to 160. The total disk capacity was now 1 MB, permitting the formatted capacity to be 744 kB. The next step in the development of the 3.5-inch disk was to change the magnetic coating thickness to 1 µm and use high-coercivity magnetic particles. The result was a linear bit density of 17,434 b/in. and a total disk capacity of 2 MB, permitting the formatted capacity of the disk to be 1.44 MB. An outline of the 3.5-inch diskette is illustrated in Figure 19–3 (Yaskawa, 1996).

THE PERSONAL COMPUTER

Beginning in the early 1960s, computer development started to evolve toward miniaturization. In 1963 the Bell Punch Company, a British firm, made the first electronic calculator from discrete transistors. It was quite large. Texas Instruments and other U.S. companies followed with smaller versions. In 1965 Digital Equipment Corporation (DEC), a fledgling manufacturer of computer test equipment, introduced the PDP-8, a limited version of a mainframe computer. It was small, made with conventional transistors, had a limited memory, and cost a fraction of the price of a mainframe. As the PDP-8 came to find applications in numerous industries and small businesses, DEC prospered handsomely. Following this success, other companies entered the minicomputer field. These machines eventually found their way into universities and technical laboratories where staff personnel, on off hours, began tinkering with programs, developing simple computer games, devising simulations of ongoing projects, drafting personal memos, and so on. This was an opportunity for hands-on computer experience, and computer interest heightened.

In 1969 a Japanese calculator manufacturer requested Intel Corporation to develop a semiconductor chip for a programmable electronic calculator. In 1971 one of Intel's engineers, after studying the Japanese request, conceived a plan for a set of general-purpose chips that could be used not only for a programmable calculator, but for most any other automatic device that could be programmed to perform a specific automatic task. Intel's general-purpose chips played an important role in the miniaturization of many automated products, including the personal computer.

In 1973 development engineers at DEC built a scaled-down version of the PDP-8, a microcomputer that came with a monitor, a keyboard, and a floppy disk drive. Floppy disks, now a standard part of computers, were new at the time, and the DEC engineers never got theirs to run properly. Digital's management was skeptical about the marketability of microcomputers and the project was abandoned. After all, what would be the use of a microcomputer?

A small firm, Micro Instrumentation Telemetry Systems (MITS), made electronic gimmicks for model rocket hobbyists. In 1974 MITS made the first personal computer using Intel's general-purpose chips and called it the Altair. The Altair was very limited in what it could do. The user wrote a program in machine code and entered it bit by bit by front panel toggle switches. At first the Altair had limited sales, since it was sold only in kit form and had to be assembled by the customer. In the January 1975 issue of *Popular Electronics* magazine, MITS published an article for electronic hobbyists on how to build a personal computer. "World's First Minicomputer Kit to Rival Commercial Models: Altair 8800" was on the magazine's cover. In spite of Altair's limitations, sales all over the country were brisk. Do-it-yourselfers began modifying their Altairs by adding extra circuitry, various accessories, and gadgets. Soon Altair computer clubs were formed, and an Altair computer fair brought computer enthusiasts together to exchange ideas. Club newsletters were published, and interest in personal computers rapidly increased.

Apple Computer, Radio Shack, and Commodore Business Machines first introduced mass-produced personal computers in 1977. Radio Shack, a chain of stores that sold specialty electronics, offered personal computer programs recorded on a plug-in cassette tape recorder. In the latter part of 1977, recognizing the direct access advantage of the flexible disk drive over the required end-to-end tape search for specific information, Radio Shack incorporated a 5.25-inch floppy disk drive attachment as an upgrade to its personal computer, the TRS-80. In 1978 Radio Shack made provisions for from one to four 8-inch flexible disk drives in the TRS-80 Model 2. Radio Shack stores were widespread and gave the Tandy Corporation, the parent company, the best outlet of any microcomputer manufacturer.

The Apple computer became very popular when it introduced the first electronic spreadsheet—a computerized accounting program. Other essential computer application programs such as word processing, for the purpose of documentation, became available, and the personal computer industry took off with many highly competitive hardware manufacturers, floppy disk manufacturers, personal-computer magazines, and computer books, as well as the establishment of stores for the sale of computers and computer accessories. Small computers with built-in monitors, keyboards, and disk drives could be purchased by individuals, were easy to use, and appealed to the general public. They also began to show up in small and medium-sized businesses and in primary and secondary schools.

In 1981 IBM introduced the Personal Computer (PC). To keep cost in line with competition, the PC was constructed from parts manufactured outside the company, with the exception of the keyboard. It was modular in. design, with plug-in components that allowed owners to upgrade. The keyboard was separate, and the monitor sat on top of the box containing the electronics. Two floppy disk drives were provided for data and program storage. A small Redmond, Washington, software company named Microsoft signed on to provide an operating system for the computer, as well as a version of the programming language, BASIC. Intel agreed to supply the multitude of computer chips. The power supplies would be furnished by Zenith Electronics, monitors would be supplied from Taiwan, the printers would be made by Epson (a Japanese firm), and a small company named Tandon Magnetics won the contract to supply 5.25-inch disks and drives. To permit programming by outside software companies to begin long before the first PC rolled off the assembly line, the details of the machine's design were revealed to the public as soon as they were finalized. The strategy was to get independent programmers to write more programming applications for IBM's personal computer than for the competition's. The strategy paid off and was a key enabler for the entire PC business of today, as was IBM's name on a PC, legitimizing the product to the business world.

The IBM PC was sold through independent retailers, just as Apple, Commodore, Tandy, and many others marketed their machines. It was a very popular computer. In 1981 about 35,000 were sold; in 1983 the company recorded sales of 800,000.

Because of its convenient size and its more rigid jacket, the 3.5-inch diskette, introduced in 1981, gradually became the industry standard for information exchange.

By 1983, personal computers became small enough to be portable. By 1990, development engineers had perfected the laptop computer, a battery-powered device that could rest in one's lap. Today's laptops house a hard disk drive, and a 3.5-inch flexible disk drive. Laptops are popular with business travelers, journalists, members of the legal profession, and many others.

Personal computers were increasingly interconnected with each other and with larger computers for the purpose of gathering, sending, and sharing information electronically. The uses of personal computers continued to multiply as the machines became more powerful and their software proliferated. In 1997, 37% of American households owned personal computers and used them for entertainment and games, education, personal finances, record keeping, scheduling and planning functions, word processing, and navigating the Internet. In business personal computers are used for accounting, inventory, business planning, financial analysis, and word processing—all the traditional business functions as well as technology applications.

Figure 19–4 compares an 8-inch diskette drive and diskette with a 3.5-inch disk drive and diskette.

Figure 19–4   The IBM 23-FD read-only diskette drive with 8-inch diskette, 81.66 kB (1971) dwarfs the standard (1990) diskette drive with 3.5-inch diskette, 1.44 MB. [*Courtesy of IBM Corporation.*]

HIGH-CAPACITY DESIGNS

Flexible disks and drives developed in the mid-1970s and 1980s used the same technology as that of the original 8-inch single- and dual-head drives, with modifications. From 1971 to 1991, flexible disk capacity increased by a factor of 17.5. During this period rotational speed improved from 90 rpm to 720 rpm and track-to-track movement was reduced from 333 ms to 5 ms. All were remarkable technical achievements. Personal computer hard disks originally in the 5–10 MB range had now advanced to several thousands of megabytes. Sound and pictures were soon common, along with video clips and software suites of 100 MB and more. Many personal computer applications required interchangeable/removable disks with capacity beyond that of the standard 3.5-inch diskette. Further flexible disk and drive performance improvements required :

The ability to mass-produce extremely thin disk coatings necessary to increase bit density.

Writing and reading on-track signals with low-cost stepper motors and associated mechanical positioning mechanisms.

Rotational stability as the disk's critical velocity was approached.

During the 1990s, performance advancements of the flexible disk and drive were made, but not without an inevitable increase in cost and complications. The conventional recording medium was improved by the application of a double-coating technique, whereby a thin recording layer with reduced-sized magnetic particles was applied over a smoothing nonmagnetic undercoat. This method helped head–media compliance, thereby decreasing reading errors due to surface roughness. Also, reducing the size of the magnetic particles permitted the recording density to be increased.

In the quest for higher rotational disk speed to increase data transfer rate, two methods were exploited to achieve disk stabilization. The first, the Bernoulli technique, used a fixed plate adjacent to the rotating disk and an air passage for air intake at the disk hub. Air impelled radially outward by the disk rotary action drew the disk toward the plate and enabled it to rotate without vibrating at a higher velocity. A slot in the plate was provided for the recording head to communicate with the disk surface. Careful design of the face of the head contributed further to disk rotation stability. The second technique employed two stationary plates, each with symmetrical opposing heads; access slots were placed on either side of the rotating disk. With this arrangement, the rotating disk was found to be free of vibrations at advanced rotational speeds.

Track-following servos that dynamically control the position of the head permit extended track densities by circumventing track runout plus environmental and stepper motor tolerances. Servo actuators used to move the head toward the center of a track are similar to those developed for loudspeaker systems (i.e., a permanent magnetic structure with a movable coil attached to the head). Current to the actuator is controlled by the track positioning error, the difference between the on- and off-track head positions. Flopticals (a contraction of floppy and optical) are magnetic flexible disks with grooves embossed between tracks. The reflection of the variations of light from the sides of the grooves into an optical system is used to generate the track positioning error. In 1991 Insite Peripherals Incorporated shipped 8-inch optical servo disks with a user capacity of 20.8 MB.

The embedded servo system is an alternate method of producing the track positioning error signal. Position signal data can be magnetically written at the factory at regular sector positions on all data tracks. When such a disk is used for reading or writing, the sector servo information is sampled and the servo position error ascertained. In 1994 the Iomega Corporation shipped Bernoulli 5.25-inch disks using a sector servo having a user capacity of 230 MB. Today's most popular flexible disk sector servo drive is the 3.5-inch Iomega ZIP™ drive, having a user capacity of 100 MB. The Iomega ZIP drive and disk first appeared in 1995.

Forecasts indicated that only 10% of the total number of flexible disk drives marketed in 1997 would be of the high-capacity or superfloppy type. Reasons cited were cost and the observation that 70 to 80% of today's personal computer applications are related to document production (a single 2 MB standard diskette holds over 3000 pages of text).

The personal computer, and its companion the floppy disk, have played an important role in our society. With further advances in technology and the development of new hardware and software, they will continue to impact our lives in many ways.

REFERENCES

Augarten, S., *Bit by Bit: An Illustrated History of Computers,* Ticknor & Fields, New York, 1984, pp. 253–280.

Capron, H. L., and B. K. Williams, *Computers and Data Processing,* 2nd ed., Benjamin/Cummins, Menlo Park, CA, 1984.

Case, J., *Digital Future: The Personal Computer Explosion—Why It's Happening and What It Means,* Morrow, New York, 1985.

Engh, J. T., "The IBM Diskette and Diskette Drive," *IBM Journal of Research & Development,* 25th anniversary issue, **25**, 701–710 (1981).

Flores, R., and H. E. Thompson, "Magnetic Record Disk Cover," U.S. Patent 3,668,658 (1972).

Fuller, F., and W. Manning, *Computers and Information Processing*, Boyd & Fraser, Danvers, MA, 1994.

Muroga, S., "Computers," in *The New Encyclopedia Britannica: Macropedia*, Vol. 16, Chicago, 1995, p. 639.

Noble, D. L., "Some Design Considerations for an Interchangeable Disk File," *IEEE Transactions on Magnetics*, **MAG-10**, 571–574 (1974).

Pugh, E. W., L. R. Johnson, and J. H. Palmer, *IBM's 360 and Early 370 Systems*, MIT Press, Cambridge, MA, 1991.

Stern, R. A., and N. Stern, *An Introduction to Computers and Information Processing*, Wiley, New York, 1982.

Time-Life Books, Eds., *The Personal Computer*, Time-Life Books, Alexandria, VA, 1989, pp. 9–18.

Yaskawa, S., S. R. Perera, and J. Heath, "Data Storage on Flexible Disks," in *Magnetic Storage Handbook*, 2nd ed., C. D. Mee and E. D. Daniel, Eds., McGraw-Hill, New York, and IEEE Press, Piscataway, NJ, 1996, pp. 3.1–3.37.

20 Instrumentation Recording on Magnetic Tape

Finn Jorgensen

Instrumentation recorders grew out of professional audio recorders in the late 1940s. The following decades witnessed fast-paced engineering improvements in instrumentation recorders that went from analog through FM (frequency modulation) to HDDR (high-density digital recording) record/playback methods. Current designs embrace the latest digital techniques and hardware from the computer data storage industry.

EARLY DATA RECORDING USING MODIFIED AUDIO RECORDERS (1935-1950)

As soon as engineers and scientists involved with measurements became aware of the magnetic tape recorder, they determined to use it to record data. A first indication of an engineering application appeared in 1936, when a recorder was used to preserve a record of the rudder deflection on an airplane during flight test.

In 1940 a recorder was used in a flight test by Lockheed. This equipment used steel tape, and two 8-channel units could be coupled to provide a total recording capability of 16 channels. The assembly was used for recording voice comments, noise, vibration, and carrier recording from strain gauge transducers. Noise and vibration recordings were subjected to frequency analysis during reproduction. Also in 1940 there appeared a German report discussing magnetic recording in connection with a phase direction finder.

During the second world war, recorders found increasing applications in recording data, and in 1945 an apparent first reference to FM carrier recording was published by Shaper, who described a loop recorder with a steel tape loop running at 25 feet per second (300 in./s), recording a 10 kHz carrier. Shaper used FM recording with a 40% deviation, achieving a 40 dB signal-to-noise ratio.

Magnetic recording found increasing applications during the late 1940s. Measurements of the forces during a parachute opening and descent were recorded by means of a system in which an FM carrier served for all data channels. One track was apparently recorded with a steady signal frequency, since flutter compensation and playback speed control were mentioned. Any speed deviation from a nominal value results in a corresponding change in signal frequency. The difference (error) signal (compared with a crystal-controlled signal) may then be FM-detected and subtracted from the similarly FM-detected signal. This technique, called flutter compensation, is widespread in instrumentation recording/playback.

The speed compensation is achieved by injecting the detected error signal into the capstan motor speed control. This technique has today reached a performance level that permits the correction of speed variations in one-thousandth of a second, hence the removal of time errors and flutter components below 1 kHz.

In 1948, L. G. Killian described a number of recording mechanisms, some with 6 and some with 12 tracks. It appears that 1948 marked the appearance of the first multichannel head with precision alignment of gaps by lapping each half of the pole-piece assembly (today called a half-block) before final assembly of the head. An early design provided 13 tracks on 1-inch-wide tape, with 50-mil track width. Heads with up to 200 tracks for a 2-inch-wide tape were made later.

THE FIRST MULTICHANNEL INSTRUMENTATION RECORDERS (1950–1980s)

Successful recording of the *Bing Crosby Show* in August 1947 resulted in an order from ABC for 12 clones of the Magnetophon, to be built by Ampex. This development paved the way for quantity production of tape recorders, first for audio, then instrumentation, and later video.

In 1950 the Naval Missile Test Station in Point Mugu, California, desperately needed a method for recording and storing the large quantity of data the facility was beginning to collect for engineering analysis. The station was receiving multiplexed data consisting of 10 carriers that were FM-modulated ±7.5% with 200 Hz data; the total bandwidth was 200 Hz to 60 kHz.

Raytheon was responsible for telemetry at Point Mugu. The company initially considered the use of the new Ampex audio recorders: 10 units would be needed, since each recorder had a bandwidth of only 15 kHz and therefore could accommodate just one channel. Then Raytheon learned of a bad experience with recording telemetry using an audio recorder at Edwards Air Force Base. Audio recorders have a large amount of record current boost above 1 kHz, to compensate for recorder losses and make up for the lower energy in voice/music above 1 kHz. Part of this boost is compensated upon playback.

Many engineers who applied audio recorders in making instrumentation measurements fell into the trap of assuming that a flat spectrum for the input signal would be satisfactory. In reality, the high-frequency boost severely overloads the record function with resulting severe distortion. The representative for Ampex in southern California, Bing Crosby Enterprises (BCE), took one of the company's Model 300 audio recorders to Point Mugu for a demonstration and helped the telemetry engineers in balancing the record input spectra to a flat level by simply reducing the high-frequency gain for each data channel.

The telemetry people at Point Mugu were also told that the bandwidth could be increased by simply increasing the tape speed and redesigning the electronics and heads. They were elated and wished to place an order for a few recorders. They started by procuring an Ampex Model 300 audio machine, beefed up to run at 60 in./s and provide a bandwidth of 100 kHz.

The next entry from BCE was the Model A, which offered a closed-loop capstan drive—a BCE innovation—with reversing idler plus shielded heads. This is believed to be the first instrumentation recorder designed from ground up, operating at 60 in./s and giving low flutter. The units were produced at Bing Crosby Enterprises on Sunset Boulevard in Hollywood, California. Point Mugu and the navy wished to have a large number of recorders. BCE accepted the order on behalf of Ampex, which went ahead and designed an instrumentation recorder to fit the need.

Ampex then declined to sell to BCE, and subsequently it was war between the two companies. Ampex canceled all arrangements with BCE, which then went on to design Model B, which ran at a speed of 360 in./s and had a signal bandwidth of 1 MHz. Model B was the first portable instrumentation recorder, weighing just under 800 pounds, in three pieces (see Fig. 20–1). The $125,000 unit was usable for airborne operation, had 400 Hz motors, and fully met military specifications. Some of the Model Bs were used to monitor broadcasts over Eastern Europe. Ampex developed Model FR600, which also use a closed-loop capstan, relying on tape reel tensions to produce tension within the capstan loop.

In 1950 M. J. Stolaroff and K. B. Boothe each published a paper describing the use of magnetic tape equipment in telemetering. This marked the beginning of large-scale instrumentation use of tape equipment in the United States. Stolaroff reported heavy use of FM/FM subcarrier recording. He also wrote "that flutter compensation should be used for recording FM telemetry subcarriers, because it would be impractical to produce low-flutter transports in quantity and, even if this were done, maintenance would require highly skilled technicians and the recorder would be put in the class of delicate, precision instruments." The tests reported by Stolaroff were conducted with Ampex equipment designed for broadcast purposes but modified to run the tape at 15 and 30 in./s, rather than 7.5 and 15 in./s.

Figure 20–1   First portable recorder, Model B, from Bing Crosby Enterprises, 1955. [*Courtesy of Jack Mullin.*]

The 3M Company acquired Bing Crosby Enterprises on August 31, 1956, and renamed it Mincom Division, 3MCo. The next instrumentation recorder in this series was the Model C, which greatly improved the capstan drive by grinding the capstan shaft to a profile that would provide differential tape tension.

The tape moving into the capstan loop has its speed controlled by the surface of the smaller radius center portions of the capstan (see Fig. 20–2). The outgoing tape has its speed controlled by the outer surfaces, which have larger radii. Tape is therefore moved faster out of the capstan loop than going in to it, and that is how the differential tension is generated. The flutter resulting from variations in reel rate and reel tension was smaller than in the FR-600, which relied on reel holdback tension to control the tape tension over the heads.

The 800-pound "portable" recorder in Figure 20–1 was more than equaled in size by the matching ground station recorders/reproducers, generally housed in 19-inch rack enclosures, 6 to 7 feet tall. The top of the rack contained a tape

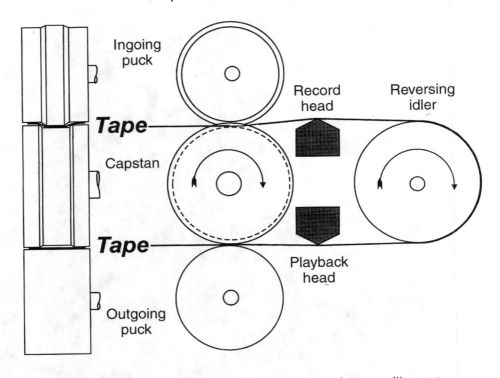

Figure 20–2   The Mincom Model C differential capstan has two radii contacting the center and the edges of the tape, thus producing tape tension.

transport that could handle up to 15-inch-diameter tape reels. Electronics and power supplies were in the bottom of the racks. Figure 20–3 shows an early BCE unit in preparation for shipment.

As reported by Athey in 1966, many companies were active in the development of instrumentation recorders. U.S. firms included Ampex, Datatape (from Bell & Howell and Astro Science), Hewlett-Packard, Honeywell, Mincom, Precision Echo, RCA, and Sangamo. European entrants were Bruel & Kjaer, BTS, EMI, RACAL, Schlumberger, Tandberg Data, and Thorn EMI. Spacecraft and crash recorders were built by Borg Warner, Fairchild Instruments, Genisco, Gould, Leach, Odetics, and RCA. Many other companies came and went over the years; the author apologizes for any omissions.

Instrumentation recording was an interesting and challenging business: the reward was often a development contract from one of several large government agencies, or their subcontractors, and a follow-on fabrication contract if the engineering prototype functioned to the agency's satisfaction. Most of the work was clever engineering rather than a buildup around a unique patented technology.

Figure 20–3   Final check out by "midnight oil" of Bing Crosby Enterprises instrumentation recorder: Dean deMoss (left), Wayne Johnson (right) aligning heads, and Jack Mullin (on break, top). [*Courtesy of Jack Mullin*.]

The optimum designs were tailored to specific applications. Rarely would a design deliver optimum performance in all areas of a specification, and compromise solutions became typical for recorders. These trade-offs led to a host of different machines that basically were analog recorders, using direct recording with different ac bias settings, FM or PCM, both with and without ac bias.

TECHNOLOGY REFINEMENTS TO INSTRUMENTATION RECORDERS USED IN SPACE TELEMETRY (1960–1990)

Recorders became a much needed device for making permanent records of varying signals on to a magnetic tape. The name *instrumentation recorder* implies a high-accuracy device for storing and replaying data of various sorts. The data were often analog voltages representing measurements, directly from transducers or from demodulated subcarriers. These were carriers that were frequency- or pulse-code-modulated.

Instrumentation recorders were required to store very large amounts of data, amounts that grew in size from year to year. The demands were reflected in the tape width increase from 0.25 inch through 0.5, 1, and 2 inches, and in tape lengths that needed up to 15-inch-diameter reels. Track densities went from 4 tracks per inch to 200 t/in., although the practical, reliable number became 14 to 28 t/in. This ceiling was in large part due to the use of reels for storing tapes, which were often exposed to a dust-filled atmosphere. A tenfold reduction in noise-produced signal dropouts occurred when tapes were enclosed in cassettes and cartridges (VHS video, 3480, and QIC tapes). Today track densities in one or two longitudinal tape transports exceed 500 t/in. This has become possible by using tracking servos, improved tape guidance, and *azimuth recording*. In the latter the record and playback gaps are not perpendicular to the direction of tape travel; rather, they deviate plus a few degrees for all odd-numbered tracks, and minus a few degrees for all even-numbered tracks.

Early recorder specifications reflected an increasing demand in terms of signal quality, and the engineers' time was spent on applying the latest technologies to recorder components. The primary goals were improvements in the following areas:

Signal quality

Signal linearity (low distortion, distinct separate signal levels)

Wideband signal spectrum

High signal-to-noise ratio (low noise, cross talk)

Low phase shifts

Small signal amplitude fluctuations

Short gap vector recording

Tape transport

Speed accuracy and low time-displacement errors

Minimum low-frequency speed variations (wow and flutter)

Minimum high-frequency speed variations (scrape flutter)

The following sections outline the effects, causes, and possible cures for these items.

Signal Linearity

Signal linearity has been achieved by using ac bias in the direct analog recording mode. The ac bias circuitry provides a high-level, high-frequency signal that is added to the data record current. Its magnitude is typically 10 (or more) times the data signal level, and its frequency 5 (or more) times the highest data frequency. The highest recorded level has a second harmonic distortion of 1% when the level is approximately one-quarter of the signal saturation level (12 dB below saturation), which leaves an 80–90 dB linear range at midfrequencies.

Amplitude Variations

Amplitude variations limit the comparable instrument accuracy of the analog recording signals. The variations are due to modulation noise and tape coating irregularities (e.g., nonuniform dispersion, a limited number of particles or grains under the read-head gap, nonuniform coating thickness, surface roughness, variations in magnetic properties). The amplitude of the read signal fluctuates as if it were amplitude-modulated with a modulation index that typically has a magnitude of ± 0.5 to ± 1.0 dB. As discussed by Jorgensen, and also by Völz, the equivalent signal accuracy can be evaluated. For a signal-to-noise-ratio of 40 dB and $m = \pm 1$ dB, the accuracy is $\pm 2\%$—only 16 discrete signal levels can be distinguished.

Wideband Signal Spectrum

Data were originally quite limited in bandwidth owing to the transducers producing the signals. A bandwidth of 10 to 20 kHz was often satisfactory. But as the number of signal sources grew, so did the total bandwidth. It became useful to let the data channels frequency-modulate separate carriers. Ten channels, each occupying ± 10 kHz, would then occupy 200 kHz. These 10 channels then could be recorded simultaneously on a single channel of that bandwidth. FM recordings have several advantages, including excellent linearity with response to dc and insensitivity to noise.

The 10 channels were often originally transmitted as a frequency-modulation signal for a high-frequency carrier, a method called FM/FM. On recorders with larger bandwidth, it became possible to record this carrier signal prior to de-modulation and filtering into 10 channels. The received carrier was modulated down to a carrier frequency of, say, 600 kHz, with all sidebands recorded onto an instrumentation recorder having a bandwidth of 1.2 MHz. The carrier and its side-bands were recorded using standard direct recording techniques (ac-biased). This approach was called predetection recording. The experimenter could then repeat-edly play back the data and optimize the detection of the overall and the individ-ual FM signals.The best results were achieved when the predetection signal was upconverted before demodulation.

The original amplitude response of 10 Hz to 20 kHz from audio had been translated into 100 Hz to 200 kHz for the first high-speed instrumentation recorder. The demand for an extended high-frequency response ascended very quickly, and within a few years a 2 MHz bandwidth at a tape speed of 120 in./s was achieved. The wider bandwidth was obtained by minimizing the causes for signal loss at short recorded wavelengths. A frequency of 2 MHz recorded at 120 in./s has a wavelength of 60 μin. Therefore to ensure a loss no greater than 2 dB, the gap length in the reproduce head core must be only some 20 μin.

The recording conditions carried over from audio had to be abandoned: a high ac bias level that penetrates through a 500–1000 μin thick magnetic coating was reduced to *partial penetration* of a 200 μin thick coating. These steps resulted in a greatly reduced *demagnetization loss*. This loss prevails when the recorded half-wavelength is shorter than the recorded thickness of the coating. Ideally one would record on a thickness that is on the order of half a wavelength thick (i.e., 30 μin. (750 nm) for 2 MHz at 120 in./s). This method, called partial penetration recording, was adapted early in instrumentation recording, as well as in digital and video recordings. This method does reduce the signal at long wavelengths, but this part of the spectrum is not used in FM/FM or pulse code modulation (digital). It is interesting that demagnetization losses set in when the recorded wavelength ap-proached half the coating thickness. Playback analysis revealed that the effective coating thickness during playback is about half the actual coating thickness.

For such short wavelength recordings the surface of a coating must be very smooth, and that has indeed been another design goal for the magnetic tape manu-facturers. Data recorders in the 1990s are all digital, and linear densities are approaching 200 Kb/in. along a track. The equivalent bit length is 5 μin, and the wavelength is 10 μin. More recently a near-ideal tape has been produced by Fuji Film with an exceptionally thin and uniform coating thickness of 4 μin. (100 nm). A short record gap length also serves to promote partial penetration recording.

Signal-to-Noise Ratio

The signal-to-noise ratio is generally specified as a single number in de-cibels. We have become accustomed to the number 90 dB associated with the audio CD-ROM. This is really a weighted number, taking into account the ear's sensitivity-versus-frequency curve, originally measured by Fletcher and Munson.

The SNR of a magnetic recorder is fundamentally very small at both low and high frequencies. Inadequate signal level and low-frequency amplifier noise limit the low-end SNR to around 60 dB. The SNR increases to 80 to 90 dB in the midfrequency range, but then decreases dramatically at high frequencies.

Audio recorders take advantage of the fact that the energy in music is low at both low and high frequencies. This permits a corresponding boost of record current at these frequencies, the so-called record equalization. A further boost is permitted to compensate for losses inherent in the ac bias record process, together with demagnetization losses, head core losses, and so on. The latter boost depends on the particular recorder, while the first is standardized.

With audio, the general nature of the signal spectrum is known. If an unknown signal spectrum is recorded, then chances are great that overload and subsequent distortion will occur at low and high frequencies. This is the domain of an instrumentation recorder: often the signal spectrum is unknown, and should be recorded flat (i.e., with no boost applied). When, however, some knowledge of the signal spectrum exists, an equalizer can be designed to enhance the low-level input signal components, a mirrorlike inverse equalizer applied during playback. This method assures the maximum SNR with minimum distortion at all frequencies.

Many instrumentation applications require simultaneous recording of multiple modulated signal carriers. A special SNR-measuring technique called *noise power ratio* (NPR), borrowed from telephone communication technology, was applied as one overall distortion measure of a properly equalized recorder/reproducer rather than trying to balance many discrete SNR measurements over a range of frequencies.

Phase Versus Frequency Response

Another measure of recorder channel fidelity is how well the phase between individual frequency components is preserved in an input signal. Most electronics circuits or building blocks that have a flat amplitude–frequency response can be shown to have a phase shift–frequency response of zero. This does not hold true for a magnetic recorder channel. As the bit length (half-wavelength) becomes smaller than the thickness of a recorded tape, the timing of the recording undergoes a delay with respect to the long bit length (wavelength) recording. This requires elaborate equalization for near-perfect correction. Figure 20–4 shows the different square-wave responses from two different recorder/reproducers.

The playback equalizer used to boost the high frequencies must also be carefully designed. A simple circuit consisting of a delay line and a differential

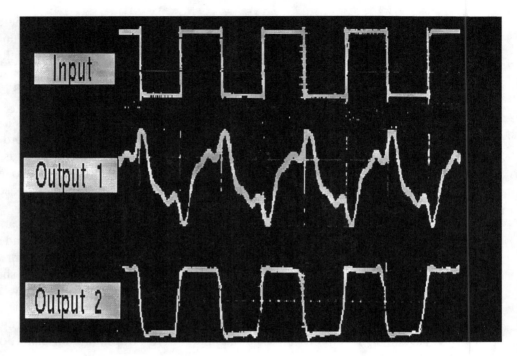

Figure 20–4   Recorder input square wave (top) and playback waveform from a reproducer with poor (middle) and good (bottom) phase equalization.

amplifier was developed by RCA for use in boosting the high-frequency energy in a video signal without inflicting phase errors on it. It was called an *aperture delay line* or *cosine equalizer*, and was first used by Mincom in the company's early instrumentation recorders. It has later found its way into digital disk drives, where it is called a *pulse-slimming equalizer*, as described by Schneider in 1975. The analog circuit gain is described by an equation that shows freedom from phase shift while boosting the high frequencies (and noise).

Precision Tape Transport

Transports for analog recording and playback were designed to provide a linear motion of the tape with as little speed variation as possible. This goal is limited by mechanical tolerances, and all transports therefore introduce some errors into the data. The speed variations are generally referred to as flutter. Their effect on data is a frequency shift that makes narrow-bandwidth analysis of recorded data difficult. Flutter also adds noise in FM systems.

The errors introduced by flutter can be quite extensive, and some recordings can be recovered only by using a transport with servo-controlled speed during playback. The recorder used in acquiring the data may have been subject to heavy vibrations and gyrations (e.g., carried in a truck, tank, aircraft, or ship).

In the early instrumentation recorder days, speed variations were corrected by recording a 17 kHz control frequency, amplitude-modulated at the rate of 60 Hz. During playback this signal was demodulated and fed into a phase comparison circuit, which was also fed with the signal from a precision 60 Hz oscillator. The circuit's output, after amplification, drove the synchronous capstan motor. The maximum error rate that could be corrected was 1 Hz.

Flutter components in a tape transport range from a few hertz to many kilohertz. The low-frequency components are generally caused by out-of-round bearings and guides, irregular tension forces from reel motors, and the like. The higher frequency components are due to a variety of mechanical vibrations that translate into the tape motion called *scrape flutter*. Therefore removal of the low-to-medium flutter frequencies does not necessarily remove all the motion errors in the tape.

The specification for speed accuracy is time displacement error (TDE). A typical value for a high-inertia transport running at 120 in./s is ±150 µs. It is possible to reduce this error to ±0.3 µs by employing a tape drive with a very low inertia capstan–motor assembly and a servo bandwidth of 500 to 1000 Hz.

Timing Errors Between Tracks

Timing errors between tracks have often hindered correlation between data on several tracks. The technical term for time differences between tracks is *skew*. Static skew that originated in tolerance buildup in a head stack may be compensated for by applying delay lines. The dynamic skew that resulted from tape flutter and irregular tracking required recorded control tones and servo control of delay lines to compensate. The amount of dynamic skew between two adjacent tracks in a head stack (assuming a tape speed of 120 in./s) was typically ±0.3 µs. It could be reduced by an elaborate servo plus variable delay-line circuitry to ± 3 ns.

Skew in today's digital instrumentation recorders is easily handled by reading all data channels into individual buffers and clocking the data out in proper timing, controlled by a phase-locked oscillator in the read electronics. Flutter and time displacement error are removed at the same time.

HIGH-DENSITY DIGITAL RECORDING (1970–CURRENT)

The many advantages of digital over analog recording changed the signal processing of instrumentation recorders. Pulse code modulation (PCM) is used in many telemetry systems, and it was natural to consider recording these data directly, by converting to saturation recording, without ac bias. The next step involved a simple electronic change to digital circuits in the instrumentation recorder. Since the chosen recorders were all in the 2 MHz/120 in./s category, they all were renamed *high-density digital recorders* or HDDRs, according to a 1986 report by Kalil and Buschman.

Digital encoding was used to extend the data bandwidth to dc, yet keep the recording free from dc. The method, called modulation coding, is distinctly different from error detection and correction coding. The choice of a suitable code was an enlightening experience for the instrumentation recorder engineers, who primarily were highly skilled analog engineers. Five leading manufacturers introduced their favorite instrumentation recorders with a new drawer-full of digital electronics, all different:

Ampex	Miller Squared (MFM^2)
Datatape	ENRZ-L
EMI	3PM
Honeywell	NRZ-L
Sangamo	RNRZ-L

This made interchange of recorded tapes impossible! The engineers had not cooperated among themselves and with their customers in choosing a joint modulation code. It took many years for the U.S. and international standard organizations, ANSI and ISO, to arrive at an industry standard, and many applications were delayed. It was also curious to learn that most modulation codes perform nearly identically, after optimization, with respect to density and other attributes. Thus N. D. Mackintosh advised in 1980: "Select a code and optimize the equipment. You will be within 7 percent performance of your competitor."

DIGITAL VIDEO TAPE RECORDERS IN INSTRUMENTATION

Data rates were limited to 2 MHz per channel in longitudinal recorders, whether in the analog or the digital mode. Higher data rates could be handled by splitting the data into several channels and recording them on separate tracks. Combination of these separate tracks is possible only if the skew between tracks is very low.

In initial instrumentation applications of analog video tape recorders, difficulties due to the switching transients between the heads in the rotating head assembly were experienced in handling analog signals. A transverse-scan instrumentation recorder/reproducer, VR-1006, was introduced by Ampex in

1961, a few years after the successful introduction of the company's MarkIV video recorder.

The first helical-scan video tape recorder for instrumentation use was produced by Echo Science in 1973, and several other companies followed suit. When digitization of data became a widespread practice, high-speed, rotating-head recorders quickly gained inroads. Later, several advances were made in improving video recorders, and their general application as data storage devices was examined.

Many rotating-head instrumentation recorders are modified broadcast recorders designed for digital video, using the D-1, D-2 and D-3 formats (Ampex, Datatape). Other rotating-head recorder formats include VHS (Metrum), DAT (Hewlett-Packard), and 8 mm (Exabyte). Figure 20–5 shows a compact (stacked reels) airborne instrumentation recorder.

The data rate in the broadcast-derived recorders has increased to 100 Mb/s, equivalent to a 50 MHz bandwidth. Higher data rates can be handled by using four channels instead of one. The incoming digital signal is divided into four channels (interleaved), and then expanded four times on the time scale. Each channel is now within the passband of the recorder. After playback, the channels are time-compressed four times and reassembled to the original data stream.

Figure 20–5   High-density, helical-scan instrumentation recorder by Datatape, with head scanner by BTS (Germany), 1982. [*Courtesy of Datatape, Inc.*]

AN ERA FADES AWAY

The use of instrumentation recorders is declining. The need for large quantities of specialized recorders for military use has diminished since the end of the Cold War. The gathering of data for intelligence agencies has also decreased. Magnetic tape recorders for spacecraft are rare, being replaced by solid state devices offering much higher reliability. Also, since all signals are now digital, specially designed instrumentation recorders are seldom necessary. Instead, any of the recorders for computer applications offer excellent performance as instrumentation recorders, from DLT, DAT, VHS, and 8 mm, and all the way down to low-cost QIC and ZIP drives.

One recent recorder casualty occurred during the upgrade of the sound and control system in Disneyland's *Tiki-Room*. All the birds' singing and motions are now controlled from a small box filled with silicon chips, replacing a vintage 1960 CM-100 instrumentation recorder.

Figure 20–6   Advanced spacecraft recorder with stacked reels. [*Courtesy of Odetics, Inc.*]

It seems proper to conclude this chapter with Figure 20–6, a photo of the most recent spacecraft instrumentation recorder, which incorporates all the best technologies accumulated over several decades, with ultraprecision mechanics and tape selected for years of trouble-free operation.

ACKNOWLEDGMENT

The author thanks Mr. Jack Mullin for several illustrations, conversations, and stories from early days to now in the field of instrumentation recording.

REFERENCES

Athey, S. W., *Magnetic Tape Recording—Technology Survey,* NASA, Washington, DC, 326 pages, 1966.

Blaupunkt-Werke, "Investigations on a Magnetic Recording Device for 20 kc," PB54174 November 1940.

Boothe, K. B., "Uses of Magnetic Tape Recording in Telemetering," *Instruments,* **23**, 1186 (1950).

Darragh, J. B., "Flight Test Data Mechanically Recorded," *Aero Digest,* **37**, 96 (1940).

Davies, G. L., *Magnetic Tape Instrumentation,* McGraw-Hill, New York, 1961.

Eldridge, D. F., and E. D. Daniel, "New Approaches to AC-Biased Recording," *IRE Transactions on Audio,* **AU-10**, 72–78 (1962).

Johnson, G. N., and W. R. Johnson, "A Predetection Recording Telemetry System," *IRE International Convention Record,* Pt. 5, 109–217 (1961).

Jorgensen, F., "Phase Equalization Is Important," *Electronic Industries,* October 1961.

Jorgensen, F., Chapter 27 in *Complete Handbook of Magnetic Recording,* 4th ed, McGraw-Hill, 1996.

Kalil, F., and A. Buschman, Eds., *High-Density Digital Recording,* NASA Reference Publication 111, NASA, Washington, DC, 1986.

Killian, L. G., "Data Recording on Magnetic Tape," *Electronic Industry & Electronic Instruments,* April 1948.

Law, E. L., "Predetection Recording of PCM Telemetry Signals," Pacific Missile Test Center, Point Mugu, CA, presented at Tape-Head Interface Committee meeting, January 1984.

Lehmann, H., "The DVL-Rudder Deflection Recorder R11/8m and Its Flight Test," *ZWB Forschungsberichte* 644 (in German), PB38249 (1936).

Lemke, J. U., "Instrumentation Recording," Chapter 9 in *Magnetic Storage Handbook,* 2nd ed., C. D. Mee and E. D. Daniel, Eds., McGraw-Hill, New York, and IEEE Press, Piscataway, NJ, 1996.

Mackintosh, N. D., "The Choice of a Recording Code," *Radio and Electronic Engineer,* **50**, 177–193 (1980).

Mallinson, J. C., Achievements in Rotary Head Magnetic Recording," *Proceedings of the IEEE,* **78**, 1004–1016 (1990).

Peshel, R. L., "The Application of Wow and Flutter Compensation Techniques to FM Magnetic Recording Systems," *IRE National Convention Record,* Pt. 7, 95–110 (1957).

Schneider, R. C., "An Improved Pulse-Slimming Method for Magnetic Recording," *IEEE Transactions on Magnetics,* **MAG-11**, 1240–1242 (1975).

Selsted, W. T., and R. H. Snyder, "Magnetic Recording—A Report on the State of the Art," *Transactions of the IRE,* September/October 1954.

Shaper, H. B., "Frequency-Modulated Magnetic Tape Transient Recorder," *Proceedings of the IRE,* **33,** 753 (1945).

Stolaroff, M. J., "Performance Results of the Ampex Magnetic Tape Recording Frequency Modulated and Pulse Width Telemetering Data," presented at the Joint *AIEE-NTF Conference on Telemetering,* 1950.

Völz, H., "Spurhöhe, Spurzahl und Kanalkapacität bei der magnetischen Speicherung," *Hochfrequenztechnik und Elektroakustik,* 1976.

Wood, R., and R. Donaldson, "The Helical-Scan Tape Recorder as a Digital Communication Channel," *IEEE Transactions on Magnetics,* **MAG-15**, 935–953 (1979).

Index

About the Editors

Eric D. Daniel has worked in magnetic recording since 1947. He spent the first 10 years with the BBC Research Department, 3 years at the National Bureau of Standards, 2 years at Ampex, and some 20 years at Memorex, where, as Director of Research, he worked on a wide variety of magnetic recording media, including computer, instrumentation, video, and audio tape products, and rigid and flexible disks. He was made a Fellow of Memorex in 1979. He retired from full-time employment in 1982.

C. Denis Mee spent some 5 years at CBS laboratories, working on audio recording, and 30 years with IBM, where he specialized in advanced storage technologies, including magneto-optical storage, magnetic recording heads, media, and recording subsystems for computer rigid disks. He was appointed an IBM Fellow in 1983. He retired from IBM in 1993, but continues to represent a consortium of companies supporting storage research at various universities.

Mark H. Clark obtained his doctorate in the history of technology from the University of Delaware in 1992 with a dissertation entitled "The Magnetic Recording Industry, 1878–1960." He is currently assistant professor of history in the Department of Humanities and Social Sciences, Oregon Institute of Technology. An authority on the early history of audio magnetic recording, he spent the summer of 1996 as a Fulbright Professor at the University of Aarhus, Denmark, researching the life of Valdemar Poulsen.